Lecture Notes in Computer Science   5329

Commenced Publication in 1973
Founding and Former Series Editors:
Gerhard Goos, Juris Hartmanis, and Jan van I

T0092552

Barbara Caputo   Markus Vincze (Eds.)

Cognitive Vision

4th International Workshop, ICVW 2008
Santorini, Greece, May 12, 2008
Revised Selected Papers

 Springer

Volume Editors

Barbara Caputo
IDIAP Research Institute, Centre Du Parc
Rue Marconi 19, 1920 Martigny, Switzerland
E-mail: bcaputo@idiap.ch

Markus Vincze
Vienna University of Technology, Automation and Control Institute
Gusshausstr. 27/376, 1040 Vienna, Austria
E-mail: vincze@acin.tuwien.ac.at

Library of Congress Control Number: Applied for

CR Subject Classification (1998): I.4, I.2.10, I.5, E.2, I.3.5-7

LNCS Sublibrary: SL 6 – Image Processing, Computer Vision, Pattern Recognition, and Graphics

ISSN   0302-9743
ISBN-10   3-540-92780-8 Springer Berlin Heidelberg New York
ISBN-13   978-3-540-92780-8 Springer Berlin Heidelberg New York

springer.com

© Springer-Verlag Berlin Heidelberg 2008
Printed in Germany

Typesetting: Camera-ready by author, data conversion by Scientific Publishing Services, Chennai, India
Printed on acid-free paper   SPIN: 12590704   06/3180   5 4 3 2 1 0

Preface

We are very pleased to present the proceedings of the 4th International Cognitive Vision Workshop, held as part of the 6th International Conference on Computer Vision Systems on Santorini, Greece during May 12–15, 2008. The aim of ICVW 2008 was to document the progress of the relatively young field of cognitive computer vision, bringing together researchers working and interested in this field and giving them a platform to discuss the results of the different European cognitive vision projects as well as international projects in this area. Original research papers were solicited in all aspects of cognitive vision, targeting the following areas in particular:

- Memory: The coupling between visual perception, tasks, knowledge and the visual system requires memory. Issues that are of special importance for integrating memory into vision systems include: how to manage representations with limited resources; model for attention; integration of information across representations and time.
- Learning and Adaptation: A system whose goal is that of interacting with the real world must be capable of learning from experience and adapting to unexpected changes. Also, there is a need for integration of multiple visual features to enable generation of stable hypotheses, and for methods for combination of cues in the presence of uncertainty.
- Categorization: Research has in particular focused on recall of specific object instances, events and actions. Whereas recently some progress has been achieved in systems that allow limited recognition of object classes, events and scenes across visual appearance, new methods are needed to enable abstractions and effective categorization across variations in color, surface markings, geometry, temporal scenes, context and tasks.
- Integration: Vision is often considered in isolation. When considered in the context of an embodied system the concept of an "active visual observer" becomes important. The visual system operates here as a task-oriented perception module that generates a diverse set of visual descriptions about the environment. The set of descriptors is by no means organized in a hierarchy. Depending on the task at hand the system might generate features to the "agent" in terms of events, labels, and/or spatio-temporal models (geometry, trajectories, relations, etc.). Thus, integration plays an important role, from processing of visual cues and multi-modal sensor fusion to systems architecture.

The 17 submission received by the Program Committee (PC) were first reviewed by three PC members each, and then advocated by a workshop organizer. Overall, 11 papers were selected for oral presentation. Authors were given two months after the workshop to revise their papers, according to the reviews and the comments received during the workshop. The 11 papers gathered in this volume cover important aspects of cognitive vision like face recognition, activity interpretation, attention, memory maps and scene interpretation.

It is our pleasure to thank all the members of the PC and the people involved in the workshop organization, in particular Dimitris Christos, who designed and updated the workshop website. Last but not least, we thank all authors who contributed to this volume for sharing their new ideas and results with the community of researchers in this rapidly developing field. We hope that you enjoy reading this volume and find inspiration for your future work in the papers contained here.

October 2008   Barbara Caputo
                                                                Markus Vincze

Organization

ICVW 2008 was organized as a satellite event of the 6th International Conference on Computer Vision Systems, that was held on Santorini, Greece on May 12–15.

Workshop Organizers

Barbara Caputo   Idiap Research Institute, Switzerland
Markus Vincze   Automation and Control Institute,
     Vienna University of Technology

Program Committee

Ales Leonardis   Darius Burschka
Bernt Schiele   Vasek Hlavac
David Hogg   Hans-Hellmut Nagel
Giorgio Metta   Antonis Argyros
Nick Pears   Luis Alberto Munoz Ubando
John Tsotsos   Wolfgang Forstner
Justus Piater   Ross Beveridge
Michael Felsberg   Horst Bischof
Sven Wachsmuth   Rod Grupen
Danijel Skocaj   Simone Frintop
Francesco Orabona   Mario Fritz

Table of Contents

Face Recognition and Activity Interpretation

Attention, Search and Maps

Scene Interpretation

Face Recognition with Biologically Motivated Boosted Features

Erez Berkovich[1], Hillel Pratt[2], and Moshe Gur[1]

[1] Biomedical Engineering Department, Technion – Israel Institute of Technology,
Haifa, 32000 Israel
`bmerez@tx.technion.ac.il, mogi@bm.technion.ac.il`
[2] Evoked Potentials Laboratory, Technion – Israel Institute of Technology,
Haifa, 32000 Israel
`hillel@tx.technion.ac.il`

Abstract. The current work presents a new face recognition algorithm based on novel biologically-motivated image features and a new learning algorithm, the Pseudo Quadratic Discriminant Classifier (PQDC). The recognition approach consists of construction of a face similarity function, which is the result of combining linear projections of the image features. In order to combine this multitude of features the AdaBoost technique is applied. The multi-category face recognition problem is reformulated as a binary classification task to enable proper boosting. The proposed recognition technique, using the Pseudo Quadratic Discriminant Classifier, successfully boosted the image features. Its performance was better than the performance of the Grayscale Eigenface and L,a,b Eigenface algorithms.

Keywords: Face Recognition, Biologically Motivated Image Features, Boosting, Pseudo Quadratic Discriminant Classifier (PQDC).

1 Introduction

The computer vision community is trying to tackle the problem of face recognition for more than 30 years [1]. Recently, there is a renewed interest in this field due to emerging homeland security needs. Currently, there are many commercial face recognition systems, and the potential for more applications is vast: public security, video tracking and monitoring, law enforcement and biometric authentication.

The academic work in this field led to significant improvements in performance in recent years [2]. However, many face recognition systems need to operate under controlled conditions of pose and illumination with cooperative subjects in order to achieve good recognition performance.

Conventional face recognition methods usually use the 2D image matrix of the human face to derive a specific representation of the subject. This image representation is inherently susceptible to changes in subject's pose and scene illumination. Good recognition methods need to be indifferent to such extrinsic changes, and should produce face representations with small within-class variation and large between-class variation.

B. Caputo and M. Vincze (Eds.): ICVW 2008, LNCS 5329, pp. 1–13, 2008.

The Eigenface model for face recognition [3] is considered the first successful example of facial recognition algorithm. It provides a representation of each face as a linear combination of eigenvectors. These eigenvectors are derived from the covariance matrix of the high-dimensional face vector space. The biometric template of each face is composed of its projected coordinates in "face space". The advantage of this model is the efficient computation of biometric templates which is ideal for identification purposes. Still, its main disadvantage is the significant decrease in recognition performance when faces are viewed with different levels of lighting or pose.

Moghaddam and Pentland [4] used a probabilistic method to model the distributions of intra-personal variation (variation in different images of the same individual) and extra-personal variation (variations in image appearance due to difference in identity). Both high-dimensional distributions are estimated using eigenvectors. This probabilistic framework is particularly advantageous in that the intra/extra density estimates explicitly characterize the differences between identities.

Other face recognition techniques apply transformation to the 2D face image in order to produce unique face representations. The Eigenphase method [5] uses Fourier transform and models the phase spectrum of face images. Principal Component Analysis (PCA) is performed in the frequency domain on the phase spectrum leading to improved recognition performance in the presence of illumination variations.

Wavelets transforms have been extensively used for face recognition tasks, the most common of which is the Gabor wavelets. The sensitivity response of the Gabor oriented filters is similar to that of orientation selective neurons in the visual cortex and exhibits desirable characteristics of orientation selectivity and spatial locality [6]. Elastic Graph Matching (EGM) extracts concise face descriptions called Gabor Jets and represents a face as a labeled graph [7]. Each vertex of the graph corresponds to a fiducial point on the face (eyes, mouth, etc.), and is depicted by a multi-orientation and scale Gabor Jet computed from the image area around the vertex landmark. The edge of the graph represents the connection between two vertices landmarks. After the construction of the graph, recognition is based on a straightforward comparison of reference and probe image graphs. This enables EGM to model local facial features as well as global face configuration. The approach is susceptible to imprecise landmarks localization and cannot learn to maximize the extra-personal to intra-personal distance.

Liu [8] introduced another Gabor based face classifier which produced excellent recognition results. The Gabor-Fisher Classifier (GFC) method derives an augmented Gabor feature vector from each of the image pixels multi-scale and orientation Gabor wavelet representation. This high dimensional representation is uniformly down-sampled and then projected to a lower space using Principal Component Analysis. Finally, the Enhanced Fisher linear discriminant Model (EFM) is used for the discrimination of the resulting face templates. Although this method was very successful in face recognition tasks, its uniform down-sampling procedure is arbitrary and can keep redundant Gabor features while discarding informative ones which can unfavorably affect the recognition process.

The AdaBoost Gabor Fisher Classifier (AGFC) [9] was proposed in order to tackle the above-mentioned problem of GFC. In AGFC, AdaBoost [10] is exploited to optimally select the most informative Gabor features. The selected low-dimensional AdaGabor features are then classified by Fisher discriminant analysis for final face identification. It was shown that the proposed method effectively reduced the dimensionality of Gabor features and that the final recognition performance has also been improved.

In both GFC and AGFC the holistic feature vector of the face image does not utilize the spatial information of human face. Additionally, these methods are based solely on the grayscale information of the face images without incorporating the color information which can enhance the recognition performance. The current work presents a new face recognition algorithm which tries to overcome the disadvantages of the previously mentioned methods, while introducing novel biologically motivated image features and a new learning algorithm, the Pseudo Quadratic Discriminant Classifier (PQDC). The recognition approach consists of constructing a face similarity function, which is the result of combining linear projections of biologically motivated image features. These projections are learned from a bi-categorial training database which is constructed from "same-person" and "different-person" image pairs.

2 Methods

2.1 Biologically Motivated Low-Level Features

The primary visual cortex contains cells tuned to collection of features that create the perceived object representation. Similarly, in the current recognition model, images are decomposed to low-level features resembling those of the primary visual pathways.

The features are produced by extending the Hunter L,a,b color space [11] to resemble the receptive fields found in the visual cortex. Hunter L,a,b color space is a perceptually linear color-opponent space with a dimension L for luminance and dimensions a and b representing red versus green and yellow versus blue color-opponency, respectively. This representation is based on nonlinearly-compressed CIE XYZ [12] color space coordinates. In this color space, which approximates human vision, a change of the same amount in a color value should produce a similar visual difference.

For computation of Hunter L,a,b color space, image pixel values are linearly transformed from the CIE RGB space to XYZ space:

$$\begin{bmatrix} X \\ Y \\ Z \end{bmatrix} = \begin{bmatrix} 0.412453 & 0.357580 & 0.180423 \\ 0.212671 & 0.715160 & 0.072169 \\ 0.019334 & 0.119193 & 0.950227 \end{bmatrix} \cdot \begin{bmatrix} R \\ G \\ B \end{bmatrix} \quad (1)$$

During the computation of the basic features the face image is decomposed to intensity, color, and orientation channels. The following extension of the Hunter L,a,b color space is introduced to produce biologically motivated features. Similarly to Itti et al. [13], spatial scales of the visual receptive fields are created using dyadic Gaussian pyramids, which progressively low-pass filter and sub-sample each of the resulting L, a, b channels. Although the proposed features are similar to those presented in [13], the introduction of the nonlinear Hunter L,a,b color space is novel to our knowledge and led to improved recognition results in comparison to other linear color spaces.

2.1.1 Intensity Channel

Intensity features are computed by linear "center-surround" operators resembling visual receptive fields. Practically, the pyramidal representation of the input is convolved with a center-surround filter to produce the pyramidal intensity channel. Adapting Itti's [13] notation we denote across-scale difference between two intensity maps as "Θ" which is obtained by interpolation to the finer scale and point by point subtraction. This extension of the luminance L channel leads to the following center-surround pyramidal representation:

$$L_j = 100\sqrt{\frac{Y_j}{Y_n} \Theta \frac{Y_{j+1}}{Y_n}}, \quad j = 1,\ldots,m-1 \tag{2}$$

Where L_j is the j-th level luminance value, Y_j is the j-th level Y tristimulus value, m is the number of pyramidal levels and Y_n is the Y tristimulus value of a specified white object.

2.1.2 Opponent-Color Channels

The Hunter opponent color axes a and b which roughly represent redness (positive) versus greenness (negative) and yellowness (positive) versus blueness (negative) are also extended. For the j-th level a and b channels we get:

$$a_j = K_a\left(\frac{X_j}{X_n} \Theta \frac{Y_{j+1}}{Y_n} \middle/ \sqrt{\frac{Y_{j+1}}{Y_n}}\right), \quad j = 1,\ldots,m-1$$

$$b_j = K_b\left(\frac{Y_j}{Y_n} \Theta \frac{Z_{j+1}}{Z_n} \middle/ \sqrt{\frac{Y_{j+1}}{Y_n}}\right), \quad j = 1,\ldots,m-1 \tag{3}$$

Where K_a, K_b are coefficients which depend upon the illuminant, X_n is the X tristimulus value of the specified white object and Z_n is the Z tristimulus value of the specified white object (for a D65 illuminant, $K_a = 172.30$, $K_b = 67.20$, $X_n = 95.02$, $Y_n = 100.00$ and $Z_n = 108.82$).

2.1.3 Orientation Channels

Local orientation information is obtained using oriented Gabor filters. Four orientations are being used: 0°, 45°, 90°, and 135°. The sensitivity response of the oriented filters is similar to that of orientation selective neurons in the visual cortex and is modeled by:

$$G(x, y) = \exp\left(-\frac{X^2 + \gamma^2 Y^2}{2\sigma^2}\right) \times \cos(2\pi\frac{X}{\lambda}) \tag{4}$$

Where $X = x \cdot \cos(\theta) + y \cdot \sin(\theta)$ and $Y = -x \cdot \sin(\theta) + y \cdot \cos(\theta)$, and the filter parameters are: orientation θ of the Gabor filter stripes, effective width of the

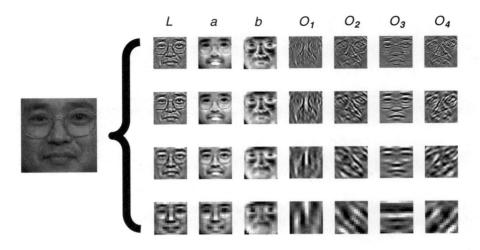

Fig. 1. The biologically motivated pyramidal image features. The columns from left to right display the output of the intensity channel L, a channel, b channel, and the four orientation channels. The rows display the different pyramidal levels of the representation, from fine resolution (top row) to a coarse one (bottom row).

filter σ, wavelength of the cosine factor λ and the spatial aspect ratio which specifies the ellipticity of the support of the Gabor function is γ.

Image decomposition to seven pyramidal channels with four spatial scales results in 28 biologically motivated image features. Each "feature" is basically an image matrix at a specific pyramidal level, processed by the appropriate receptive-field sensitivity response. Fig. 1 displays the resulting biologically motivated pyramidal image features for a general input image.

2.2 AdaBoost (Adaptive Boosting)

Boosting has been proposed to improve the accuracy of any given learning algorithm. It is based on the observation that it is much easier to find many rough rules of thumb than to find a single highly accurate decision rule. In the Boosting procedure a classifier with classification performance on the training set greater than an average performance is constructed, after which new component classifiers are added to form an ensemble whose joint decision rule has an arbitrarily high accuracy on the training set.

AdaBoost (Adaptive Boosting) [10] is a specific Boosting approach. The training methodology of AdaBoost is the following: Each training example is assigned a weight that determines its probability of being selected for some individual weak classifier h_t. With every iteration, the training data are used to construct weak classifiers on the training data. The best weak classifier is weighted by its accuracy and then added to the final strong classifier. At the successive iteration, the training data is re-weighted; examples that are misclassified gain weight and examples that are classified correctly lose weight. This process is iterated until a predefined error rate is reached or enough weak classifiers have been constructed. In this way, AdaBoost

Given $(x_1, y_1), \ldots, (x_m, y_m)$ where $x_i \in X$, $y_i \in Y = \{-1, +1\}$

Initialize $D_1(i) = 1/m$

For $t = 1, \ldots, T$:

- Train weak learner using distribution D_t
- Get weak classifier $h_t : X \to R$
- Choose $\alpha_t \in R$
- Update : $D_{t+1}(i) = \dfrac{D_t(i) \cdot \exp(-\alpha_t\, y_i h_t(x_i\,))}{Z_t}$

Where Z_t is a normalization factor

Output the final classifier :

$$H(x) = \text{sign}\left(\sum_{t=1}^{T} \alpha_t h_t(x) \right)$$

Fig. 2. The AdaBoost Algorithm

ignores easy training patterns and focuses on the difficult informative patterns. A pseudocode for AdaBoost is given in Fig. 2.

2.2.1 Face Recognition as a Binary Classification Problem

The excellent performance of AdaBoost in various recognition problems, led us to apply it to the resulting multitude of image features. Although face recognition is a multi-category problem, it can be formulated as a binary classification task. Adopting the methodology of Moghaddam and Pentland [4] we label the difference feature vector of two face images as "intra-personal" if they belong to the same face and as "extra-personal" otherwise. This transforms the face recognition problem to a binary classification problem; one only has to decide whether a difference feature vector between the probe image and a gallery image belongs to the "intra-personal" or "extra-personal" classes.

The pre-processing phase produces a multitude of features describing the input image. In contrast to other recognition methods which are specifically based on pixel values, we would like to combine these features to produce better recognition results. AdaBoost is used to find the optimal combination of the biological features; given an image training set, the "intra-personal" / "extra-personal" difference classes are computed for each feature, and then AdaBoost is trained on them to select a weak classifier built upon the most informative feature. The process is iterated until a final strong classifier is constructed.

Given two images i and j, and a specific image feature k the output of a weak classifier based on the feature difference is $h_t(x_{ki} - x_{kj})$. We seek an implementation for these weak classifiers which will produce a computationally efficient strong classifier. It would be advantageous to apply classifiers which use linear transformations,

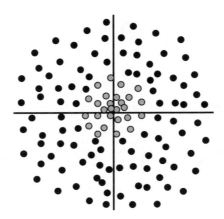

Fig. 3. A simplified visualization of the intra-personal class (gray circles) and the extra-personal class (black circles)

because the additivity condition will enable to compute feature difference in low-dimensional space instead of in the original high-dimensional pixel space: $h_t(x_{ki} - x_{kj})$ = $h_t(x_{ki}) - h_t(x_{kj})$. However, it is expected that the intra-personal class will be scattered around the coordinate origin, while the "noisier" extra-personal class will be scattered more distantly from the origin. A simplified visualization of this situation is demonstrated in Fig. 3. Common linear classifiers such as the Fisher Linear Discriminant (FLD) [14] will find it difficult to discriminate between both classes, and we therefore introduce the Pseudo Quadratic Discriminant Classifier (PQDC) which is adapted to this classification problem.

2.3 Boosting with the Pseudo Quadratic Discriminant Classifier (PQDC)

A novel discriminant method, Pseudo Quadratic Discriminant Classifier (PQDC), is proposed to discriminate between two different distributions. Suppose that we have a set of n d-dimensional samples $x_1,...,x_n$, n_1 samples in the subset D_1 labeled ω_1 and n_2 samples in the subset D_2 labeled ω_2. Similarly to Fisher Linear Discriminant (FLD) [14] we reduce the dimensionality of the data by devising a transform from d dimensions to one dimension which well separates the samples.

To overcome the fact that FLD can yield singular results for the "intra-personal" / "extra-personal" class distributions, PQDC introduces a non linear transform. This transform is composed of a linear projection vector, W, after which a non-linear quadratic operation, $()^2$, is performed:

$$y = f(x) = \left(W^T x\right)^2 = W^T x x^T W \tag{5}$$

This yields a corresponding set of n samples $y_1,...,y_n$ divided into the subsets Y_1 and Y_2. Similarly to FLD, we seek a linear projection vector, W, which well separates the subsets and maximizes the following functional $J(W)$:

$$J(W) = \frac{|\tilde{m}_1 - \tilde{m}_2|^2}{\tilde{S}_1^2 + \tilde{S}_2^2} \qquad (6)$$

Where $\tilde{m}_i$ is the projected class mean and $\tilde{S}_i^2$ is the projected class scatter.

To find the optimal projection vector, W, which best separates the distributions, we need to expand $\tilde{m}_i$ and $\tilde{S}_i$. The projected class mean is given by:

$$\tilde{m}_i = \frac{1}{n_i} \sum_{y_j \in Y_i} y_j = \frac{1}{n_i} \sum_{x_j \in D_i} \left(W^T x_j\right)^2 = \frac{1}{n_i} \sum_{x_j \in D_i} W^T x_j x_j^T W = W^T S_{B_i} W \qquad (7)$$

Where n_i is the size of class D_i and S_{B_i} is defined as:

$$S_{B_i} \equiv \frac{1}{n_i} \sum_{x_j \in D_i} x_j x_j^T \qquad (8)$$

The squared difference of the projected subsets means is a measure of the class separation. It follows that this between-class sum of squares is:

$$(\tilde{m}_1 - \tilde{m}_2)^2 = \left[W^T \left(S_{B_1} - S_{B_2}\right)W\right]^2 \qquad (9)$$

The within-class sum of squares gives the projected class scatter:

$$\tilde{S}_i^2 = \sum_{y_j \in Y_i} (y_j - \tilde{m}_i)^2 = \sum_{x_j \in D_i} \left(W^T x_j x_j^T W - W^T S_{B_i} W\right)^2 = \sum_{x_j \in D_i} \left[W^T \left(x_j x_j^T - S_{B_i}\right)W\right]^2 \qquad (10)$$

In order to simplify the expressions for the between-class scatter and the within-class scatter, we introduce $\mathbf{W}$ which is an orthogonal projection matrix. Replacing the projection vector W with the matrix $\mathbf{W}$ will produce the expression $\mathbf{W}^T S_B \mathbf{W}$ as a correlate of between-class scatter and the expression $\mathbf{W}^T S_W \mathbf{W}$ as a correlate of the total within-class scatter. In the above we defined S_B as:

$$S_B \equiv \left(S_{B_1} - S_{B_2}\right)\left(S_{B_1} - S_{B_2}\right)^T \qquad (11)$$

And S_W is defined as:

$$S_W = S_{W_1} + S_{W_2}$$
$$S_{W_i} \equiv \sum_{x_j \in D_i} \left(x_j x_j^T - S_{B_i}\right)\left(x_j x_j^T - S_{B_i}\right)^T \qquad (12)$$

The resulting criterion function $J(\cdot)$, symbolizing the between-class scatter to the within class scatter ratio can be written as:

$$\Rightarrow J(\mathbf{W}) = \frac{\mathbf{W}^T S_B \mathbf{W}}{\mathbf{W}^T S_W \mathbf{W}} \qquad (13)$$

This leads to a generalized eigenvalue problem. It is easy to show that the eigen-vector W_i maximizing $J(\cdot)$ must satisfy:

$$S_B W_i = \lambda S_W W_i \qquad (14)$$

The eigen-vector W_i can be found by solving:

$$\Rightarrow S_W^{-1} S_B W_i = \lambda W_i \qquad (15)$$

We would like to reduce the d-dimensional classification problem to a one-dimensional problem. The projection matrix $\mathbf{W}$ is composed of d eigenvectors. Therefore, we choose W_1, the eigenvector with the largest eigenvalue as the projection vector for our Pseudo Quadratic Discriminant Classifier (in the sense described by equation 5).

For a probe image feature x_i and a gallery image feature x_j the Pseudo Quadratic Discriminant Classifier can be easily evaluated (projection of gallery image features can be done offline to save computation time) and the resulting strong classifier is given by equation 16 in which w_{ti} is the projection of an i-th image feature by t-th weak PQDC classifier. b_t is the decision threshold for the t-th classifier, i.e., the point along the one-dimensional subspace separating the projected points. It can be found by smoothing the projected data and finding the point where the posteriors are equal.

The strong classifier output is used as a similarity Grade. Images belonging to the same person will produce higher grades, while images belonging to different persons will produce lower grades.

$$H(x_i, x_j) = \sum_{t=1}^{T} \alpha_t \operatorname{sign}\left[\left\{W_t^T \left(x_{ti} - x_{tj}\right)\right\}^2 + b_t\right] = \sum_{t=1}^{T} \alpha_t \operatorname{sign}\left[\left(w_{ti} - w_{tj}\right)^2 + b_t\right] \qquad (16)$$

3 Results

We have tested the proposed face recognition algorithm on the Color FERET database [15, 16]. This database contains a large variety of faces and was prepared by the American national institute of standards to advance face recognition research.

Images of 100 subjects from sets FA and FB were used for training. The FA and FB images of the same person vary only in expression (neutral versus smiling). When testing, FA images are used as gallery images and FB images are used as probes.

All images were normalized for translation, scale and rotation according to manually labeled eye positions supplied with the FERET database. The images were

Fig. 4. A sample of a normalized FERET images

scaled to a size of 128x128 pixels (Fig. 4) and their intensities were globally normalized to have zero mean and unit standard deviation.

Five different recognition methods were compared:

1) Gray-Eigenface – the dimensionality of the original intensity channel was reduced by PCA. 120 eigenvectors representing 98% of the data variance were used. The mahalanobis-cosine distance in PCA space, which was previously shown to be superior to other distance measures [17], was applied to reflect the similarity between face templates.

2) L,a,b-Eigenface - the dimensionality of L,a,b intensity and opponent-color channels was reduced by PCA. 120 eigenvectors representing 85% of the channels variance were used, and the mahalanobis-cosine distance was applied independently on each channel. The three resulting similarity measures were fused by arithmetical mean.

3) Gray-BPQDC - the original intensity channel AdaBoosted using PQDC. Different image pairs of the 100 subjects were used as training data. AdaBoost was applied for a total of 120 rounds yielding a final strong classifier consisting of 120 weak classifiers.

4) L,a,b-BPQDC - L,a,b intensity and opponent-color channels AdaBoosted using PQDC for a total of 120 rounds.

5) L,a,b-BIO-BPQDC - proposed recognition algorithm: the complete ensemble of biological image features shaped by the appropriate filters at various scales and orientations. The features were decomposed to non-overlapping blocks sized 16 by 16 pixels and then fed to the classifier. This enabled a global representation of faces at coarser pyramidal scales and a more localized representation at finer pyramidal scales. Features were AdaBoosted using PQDC for a total of 120 rounds.

The recognition performance was evaluated using receiver operating characteristic (ROC) curves. A ROC curve illustrates over all possible decision thresholds the fraction of test image pairs correctly classified as different (extra-personal) versus the fraction of image pairs incorrectly classified as similar (intra-personal). The performance of the five recognition methods, estimated from the area under the ROC graph is given in Table 1.

Table 1. Recognition performance of the five recognition methods

Method	Recognition Rate
Gray-Eigenface	0.98276
L,a,b-Eigenface	0.99068
Gray-BPQDC	0.99326
L,a,b-BPQDC	0.99873
L,a,b-BIO-BPQDC	0.99954

Worst recognition results (0.98276) were obtained for the Gray-Eigenface method, where the images were processed as grayscale; Gray-BPQDC which boosted these grayscale intensities yielded an improved result (0.99326); The performance of

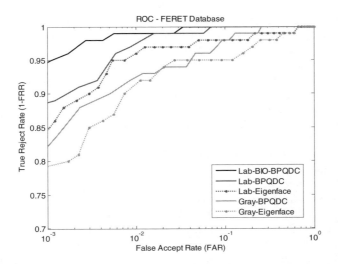

Fig. 5. ROC graph of the recognition results on the FERET database. Gray-Eigenface) Grayscale Eigenface method using intensity channel; Gray-BPQDC) Intensity channel AdaBoosted using PQDC; L,a,b-Eigenface) L,a,b Eigenface method in which images are represented in the L,a,b color-space; L,a,b-BPQDC) L,a,b intensity and opponent-color channels AdaBoosted using PQDC; L,a,b-BIO-BPQDC) The ensemble of biological features boosted using PQDC.

L,a,b-Eigenface method, which added the two opponent-color channels (a, b) to the intensity channel, was also superior (0.99068) to that of Gray-Eigenface; still, a better result was achieved for L,a,b-BPQDC which boosted the same features. The L,a,b-BIO-BPQDC method was superior to all methods and resulted in the best recognition performance (0.99954). The ROC graph for these methods is presented in Fig. 5.

4 Discussion and Conclusion

In this work we introduced novel biologically motivated image features and used them for face recognition. In order to combine this multitude of features we applied the AdaBoost technique. The multi-category face recognition problem was reformulated as a binary classification task to enable proper boosting.

The newly introduced biologically motivated image features consisted of an extension to the *Hunter L,a,b* color space, which is a nonlinearly-compressed CIE XYZ color space. This extension provided a pyramidal representation of the images, with characteristic features whose spatial responses resembled those of the primary visual cortex. In addition to their luminance, images where represented by two opponent-color channels. This greatly improved the recognition performance and leads to the conclusion that color have an important rule in face recognition.

It was shown that using the Pseudo Quadratic Discriminant Classifier (PQDC) as a weak classifier successfully boosted the performance of the features used for the face recognition task. PQDC is a non-linear modification of Fisher Linear Discriminant analysis, and it is particularly suitable for the face recognition problem when regarded as a binary classification task. Unlike other non-linear classification methods, for example Kernel-PCA [18], which does not allow for an efficient computation of the difference feature vectors, PQDC first applies a linear projection which efficiently lowers the dimension of the difference vector and facilitates the offline computation of the gallery image templates. In contrast to [9] where images are processed locally and the global spatial information of human face is not utilized, the current method applies PQDC to the ensemble of pixels of each biological feature and hence spatial information of the face is utilized.

The performance of the proposed recognition technique, L,a,b-BIO-BPQDC, was compared to other recognition methods: 1) Gray-Eigenface, based on dimensionality reduction of the intensity channel, 2) L,a,b-Eigenface based on dimensionality reduction of the intensity and opponent-color channels, 3) Gray-BPQDC - the original intensity channel AdaBoosted using PQDC, 4) L,a,b-BPQDC - L,a,b intensity and opponent-color channels AdaBoosted using PQDC, and was superior to all of them.

Our future work would focus on analyzing the properties of the most influential image features. Furthermore, we will seek an efficient implementation for the Pseudo Quadratic Discriminant Classifier and will examine larger face data sets.

Acknowledgments

Portions of the research in this paper use the FERET database of facial images collected under the FERET program, sponsored by the DOD Counterdrug Technology Development Program Office.

The authors would like to thank the reviewers for their valuable comments.

References

1. Kanade, T.: Computer Recognition of Human Faces. Birkhäuser Verlag, Stuttgart (1977)
2. Phillips, P.J., Flynn, P.J., Scruggs, T., Bowyer, K.W.: Overview of the Face Recognition Grand Challenge. In: IEEE CVPR 2005, vol. 1, pp. 947–954 (2005)
3. Turk, M.A., Pentland, A.P.: Face Recognition Using Eigenfaces. In: Proceedings of the IEEE Conference on Computer Vision and Pattern Recognition, pp. 586–591 (1991)
4. Moghaddam, B., Pentland, A.: Beyond Euclidean Eigenspaces: Bayesian Matching for Visual Recognition, Mitsubishi Electric Research Laboratories (1998)
5. Savvides, M., Kumar, B.V.K.V., Khosla, P.K.: Eigenphases vs Eigenfaces. In: Proceedings of ICPR 2004, August 23–26, vol. 3, pp. 810–813 (2004)
6. Daugman, J.G.: Uncertainty Relation for Resolution in Space, Spatial Frequency, and Orientation Optimized by Two-Dimensional Visual Cortical Filters. J. Optical Soc. Amer. 2(7), 1,160–1,169 (1985)
7. Wiskott, L., Fellous, J.M., Kruger, N., Malsburg, C.v.d.: Face Recognition by Elastic Bunch Graph Matching. IEEE Trans. On PAMI 19(7), 775–779 (1997)
8. Liu, C., Wechsler, H.: Gabor Feature Based Classification Using the Enhanced Fisher Linear Discriminant Model for Face Recognition. IEEE Trans. Image Processing 11(4), 467–476 (2002)

9. Shan, S., Yang, P., Chen, X., Gao, W.: AdaBoost Gabor Fisher Classifier for Face Recognition. In: Zhao, W., Gong, S., Tang, X. (eds.) AMFG 2005. LNCS, vol. 3723, pp. 279–292. Springer, Heidelberg (2005)
10. Schapire, R.E.: The boosting approach to machine learning: An overview. In: Denison, D.D., Hansen, M.H., Holmes, C., Mallick, B., Yu, B. (eds.) Nonlinear Estimation and Classification. Springer, Heidelberg (2003)
11. HunterLab App. Note: Hunter L, a, b Color Scale, August 1-15, Vol. 8(9) (1996)
12. Wyszecki, G., Stiles, W.S.: Color Science - Concepts and Methods, Quantitative Data and Formulae, 2nd edn. Wiley-Interscience, New York (2000)
13. Itti, L., Koch, C., Niebur, E.: A Model of Saliency-Based Visual Attention for Rapid Scene Analysis. IEEE Trans. Patt. Anal. Mach. Intell. 20(11) (November 1998)
14. Duda, R.O., Hart, P.E., Stork, D.G.: Pattern Classification, 2nd edn. John Wiley & Sons Inc., Chichester (2001)
15. Phillips, P.J., Wechsler, H., Huang, J., Rauss, P.: The FERET database and evaluation procedure for face recognition algorithms. Image and Vision Computing J 16(5), 295–306 (1998)
16. Phillips, P.J., Moon, H., Rizvi, S.A., Rauss, P.J.: The FERET Evaluation Methodology for Face Recognition Algorithms. IEEE Trans. Pattern Analysis and Machine Intelligence 22, 1090–1104 (2000)
17. Beveridge, R., Bolme, D., Teixeira, M., Draper, B.: The CSU face identification evaluation system user's guide: Version 5, Tech. Rep., CSU (May 2003)
18. Schölkopf, B., Smola, A., Müller, K.-R.: Nonlinear Component Analysis as a Kernel Eigenvalue Problem. Neural Computation 10(5), 1299–1319 (1998)

A New Method for Synthetic Face Generation Using Spline Curves

Ali Borji

School of Cognitive Sciences,
Institute for Research in Fundamental Sciences, Tehran, Iran
borji@ipm.ir

Abstract. Faces are complex and important visual stimuli for humans and are subject to many psychophysical and computational studies. A new parametric method for generating synthetic faces is proposed in this study. Two separate programs, one in Delphi 2005 programming environment and another in MATLAB is developed to sample real faces and generating synthetic faces respectively. The user can choose to utilize default configurations or to customize specific configurations to generate a set of synthetic faces. Head-shape and inner-hairline is sampled in a polar coordinate frame, located at the center of line connecting two eyes at 16 and 9 equ-angular positions. Three separate frames are placed at the left eyes center, nose tip and lips to sample them with 20, 30 and 44 angular points respectively. Eyebrows are sampled with 8 points in eye coordinate systems. Augmenting vectors representing these features and their distance from the origin generates a vector of size 95. For synthesized face, intermediate points are generated using spline curves and the whole image is then band pass filtered. Two experiments are designed to show that the set of generated synthetic faces match very well with their equivalent real faces.

Keywords: Synthetic faces, Face recognition, Face perception, Face space, Spline curves, Pattern recognition, Pattern discrimination, Psychophysics.

1 Introduction

Faces are among the most important visual stimuli we perceive, informing us not only about a person's identity, but also about their mood, sex, age and direction of gaze. Humans have a notable ability to discriminate, to recognize, and to memorize faces. Our ability to identify one another is vital to successful navigation in the community, and faces regardless of sharing the same basic features in the same basic configurations provide as a key source of person recognition. Attempts to elucidate this capability have motivated the development of numerous empirical and methodological techniques in the fields of psychology, neuroscience, and computer science. Neuroscientists and psychologists are concerned with the mechanisms underlying human face recognition. Computer scientists' goal is to automate the process for applied reasons. Face recognition systems are progressively becoming

B. Caputo and M. Vincze (Eds.): ICVW 2008, LNCS 5329, pp. 14–23, 2008.

popular as means of extracting biometric information. Face images are the only biometric information available in some legacy databases like international terrorist watch-lists and can be acquired even without subject's cooperation.

Although automatic face-recognition systems need not be forced to mimic human brain's processes, advances in brain's technique for face recognition have proved to be useful.

Several formal models of the representation, classification and recognition of artificial stimuli have been developed, which assume that the relevant stimuli are represented within a multidimensional space. The central assumptions of many of the models are closely related. This formal approach has been highly successful in accounting for human performance in laboratory experiments. In order to develop and test a formal model it is necessary to identify and control the relevant features or dimensions. The approach has, therefore, concerned the processing of sets of highly artificial and relatively simple stimuli [1]. Also Schematic faces have been used in the experiments. In [2], Brunswik and Reiter were the first to employ simplified face stimuli in a psychological study, and a recent neurophysiological study of inferotemporal neurons in macaques employed Brunswik faces to study categorization [3]. These faces are extreme face schematics (single horizontal line for mouth, single vertical line for nose within an ellipse, etc.) and are far too abstract to capture significant information about individual faces. Synthetic faces combine simplicity and low dimensional description with sufficient realism to permit individual identification [4].

Other studies of face perception have used photographs, computer averages of several photographs, or reconstructions from laser scanned faces [5]. While this has provided researchers to investigate different aspects of the topic, reliance on these stimuli has resulted in a number of important limitations.

Photographs are basically uncontrolled stimuli, which rarely match for color, size, orientation, texture, or lighting conditions. Additionally, they do not provide a systematic way of modifying face-specific image properties, which severely limits the extent to which similarities between stimuli can be measured, controlled, or manipulated. However the complexity of these stimuli has made it difficult to relate perception to the responses of underlying neural mechanisms.

It has been proposed that faces are represented as undifferentiated whole shapes, with little or no explicit representation of face parts. However, humans can also recognize a face on the basis of isolated features presented independently of the facial context or within a different context (e.g. scrambled faces), albeit with some loss of accuracy [6]. It appears then that both feature based and holistic representations can be used in face discrimination and their dependence or independence has been a matter of debate [7], [8].

To avoid some of these obstacles, we attempted to complete and optimize Wilson's method [4]. Here we added the facial features instead of generic features used by Wilson and designed a new 95 dimensional face stimulus set. It is also possible to sample a specific feature with more detail by inhomogeneous sampling in different locations. This new stimulus set provides face space components such as mean face, identity levels, and caricatures. It is also possible to morph two different faces to each other.

2 Face Sampling

We first describe how we prepared a set of digital photos. In next section, the way we generate synthetic faces from these real faces will be presented. A data set comprising photos of 110 people, half from each gender was created. Each person was photographed at a distance of 1 m in frontal view. Each Person's eyes remained straight ahead within their head in the course of photographing. People were required to have a neutral expression and emotional state. We only considered male persons without facial hairs. Everybody wearing glasses was required to remove it before photographing. Color, luminance, and contrast of all images then were adjusted using Adobe PhotoShop Me 7.0. Extra parts of all images (e.g., neck, collar, etc.) were cropped. An example of such a face after pre-processing is illustrated in Figure 1.

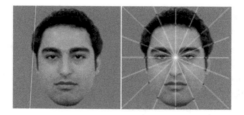

Fig. 1. A real and sampled face in outer head shape

A program was developed in Delphi 2005 programming environment (Figure 3) for sampling faces. A polar coordinate system was manually positioned to be on the middle of a line connecting the pupils of eyes. As it is shown by radial lines in Figure 1, the outer shape of the head was sampled at 16 radial axes equally spaced at polar angles of 22.5 (degree). Likewise, the inner hair was sampled by 9 further radii on and above horizontal line of the central polar coordinate system. A parameter (Degree) which controls the number of radial axes is available by changing the degree between them. For each facial feature, a separate polar coordinate system was positioned in the center of that feature.

For the left eye, center of the pupil was considered as the center of its coordinate system. It was then sampled at twenty radial axes with 18 degrees apart. Because eyes contain more spatial information along 0 and 180 axes compared with other axes we devised a mechanism for sampling around those points with more resolution. For that angles was multiplied by a scale factor greater than 1, which controls the density of angles around 0 and 180 degrees. The inverse coding mechanism was used when generating synthetic faces. Figure 2 shows a scaled coordinate frame placed at the center of the left eye.

The diameter of the iris and the thickness of the eyebrow were coded by the user subjectively in a range between 1 and 5.

The left eyebrow was sampled in the same coordinate frame located at the left eye's center. This feature was sampled in 15 degrees starting from 30 to 135.

We assumed for simplicity that the right eye and eyebrow are the mirror image of their left counterparts, so, their information were calculated as the mirror and translated code of the left eye and eyebrow.

Fig. 2. Left eye coordinate frame

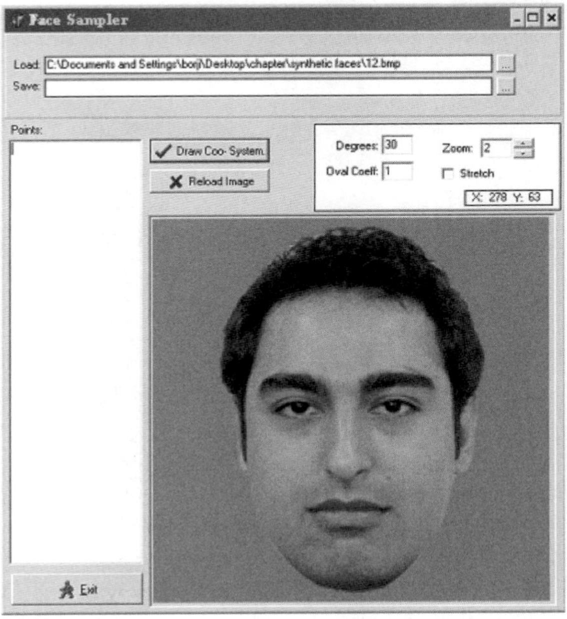

Fig. 3. Delphi program for face sampling

Another coordinate frame was positioned at the tip of the nose and was then sampled in 22.5 degrees starting from -45 to 67.5 degree. The right side of the nose was then generated by a mirror and translation operation on the information of the left side. To code the nose opening, its position was determined on 22.5, -45, or -67.5 degree.

A coordinate frame was placed in the middle of the line separating top and bottom lips. It was then sampled at 15 degrees. Finally, combination of vectors generated in each feature coordinate frame resulted in a 95 dimensional vector for each face. Figure 3 shows a snapshot of face sampling program.

3 Synthetic Face Generation

Ninety-five dimensional vector of each face was feed to a separate MATLAB file (Face Synthesizer) to generate the final synthetic faces. Main strategy here was interpolation of middle point using spline curves. Figure 4 illustrates an example of the interpolation of points of outer head shape. The same procedure was carried out for all facial features. Code snippet for interpolating outer head shape is shown below:

```
% head shape -----------------------------------------
res = 0.01;  % 2*pi/360;
y1 = 0:0.3927:2*pi+0.3927; % outer head shape
D1 = Data (5:20);  % outer head shape sampled points
cs = spline (y1, D1);
yy = 0: res: 2*pi+res;
h=polar (yy, ppval (cs,yy),'k-')
cs = spline (y1, [D2(2:11)  D1(11:16)  D2(2)]);
```

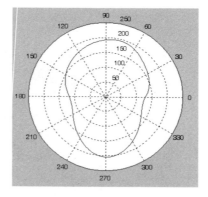

Fig. 4. Spline interpolation of outer head shape

Synthetic faces derived up to now, still lack similarity with real faces. Particularly, face parts do not have relative contrast as it could be seen in real faces. Almost in every real face, hairs and lips are usually darker than the skin, the sclera and iris of the eye are respectively lighter and darker than the skin. So, each part was filled with appropriate color to have real face-like contrast.

As face discrimination is optimal within a 2.0 octave (at half amplitude), a bandpass filtering was done with a bandwidth filter centered upon 8-13 cycles per face width [9], [10], [11], [12], [13]. Particularly, a radial asymmetric filter with a peak frequency of 10.0 cycles per face with a 2.0 octave bandwidth described by a difference of Gaussian (DOG):

$$DOG(R) = 1.26\exp(-\frac{R^2}{\sigma^2}) - 0.26\exp(-\frac{R^2}{(2.2\sigma)^2}) \tag{1}$$

where R is radius and σ was chosen so that the peak spatial frequency would be 10.0 cycles per face width on average [4]. Output of the bandpass filtering on the mid-level synthesized face is shown in Figure 5.

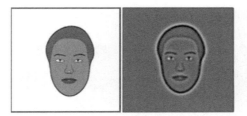

Fig. 5. Mid-level and final bandpass filtered synthetic face

Representation of faces in vector form allows algebraic operations such as morphing faces to each other, generating caricature faces, etc. Figure 6 shows a linear interpolation between two sample faces in 10 percent increments as follows:

$$C = (1 - \frac{x}{100})A + (\frac{x}{100})B \tag{2}$$

where C is the morphed face, A and B are two faces to be morphed to each other and *x* is the percent of morphing.

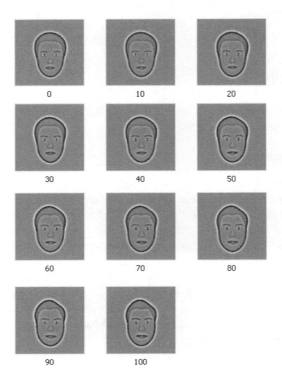

Fig. 6. Morphing two faces to each other

Some other examples of generated synthetic faces are shown in figure below:

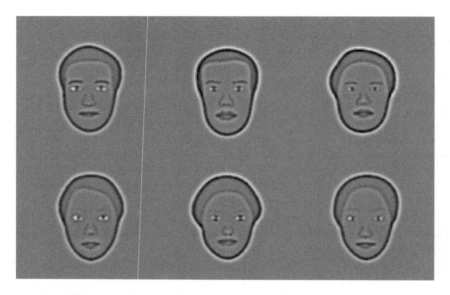

Fig. 7. Three sample synthetic male (top-row) and female faces (bottom-row)

4 Experiments

To show that our generated synthetic faces carry the major geometric information to discriminate faces, we designed two experiments:

4.1 Face Similarity Tasks

The first experiment consists of two parts. In the first part, in all trials a forced choice procedure was used with no limitation on viewing time. Each experimental run, consisted of a total number of 800 trials, and was initiated by a button press. Target image was shown in the center of the screen, and 4 alternative options were shown simultaneously besides the target and the subject's task was to match the target with its relevant option.

Experiment one, part one evaluated the similarity of synthetic facial features with their original image. As shown in Figure 8, the target was always a photograph. In half of the trails, inner facial features, and in the other half, the outer facial features were used as answer options, randomly. The subject's task was to match the most similar option with the target image.

Results showed that subjects could match both inner and outer facial features precisely with their real images. Mean performance were 75% and 86% correct for

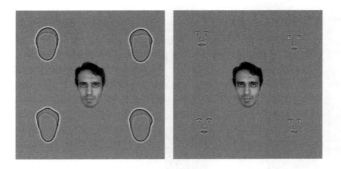

Fig. 8. Experiment one, part one test screen

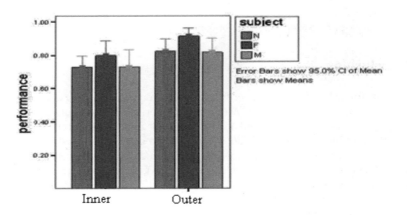

Fig. 9. Mean subject performance in first experiment

inner and outer facial features, respectively. There was no significant different between 3 subjects performing these tasks (P>0.05). Results are shown in Figure 9.

Results showed that subjects could match both inner and outer facial features precisely with their real counterpart image.

In the second part, expectedly, performance was much better when the task was matching real faces to their complete synthetic images (with both inner features and outer head shape), as shown in previous studies [4].

4.2 Gender Discrimination Task

We designed a gender discrimination task to see whether our faces carry information needed for gender categorization. In each trial, subjects were asked to determine the face gender by pressing a button indicating male or female gender.

As shown in Figure 10, mean subjects performance in this task was 0.92 and differences between female and male targets was not significant. The subjects had no significant difference as well (P>0.05).

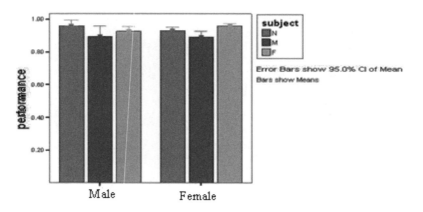

Fig. 10. Mean subject performance in gender discrimination task

5 Conclusions

A Toolbox for generating synthetic faces was introduced in this paper. It could be used for verifying both face recognition algorithms as well as conducting psychophysics tasks to understand face perception and recognition in humans and monkeys. Sampling is done with an executable program developed in Delphi 2005 and a vector of size 95 is generated which codes much detail of a real face. It is then fed to a MATLAB file, which generates a synthetic face. A set of real faces and their corresponding synthetics faces is also available and could be asked from authors. In our experiments, for the present, synthetic faces defined in a 95-dimensional face space catch very much of the information in real faces and are useful for research on face perception, memory, and recognition based upon salient geometric information in the most significant spatial frequency band.

References

1. Valentine, T.: Face-Space Models of Face Recognition, Computational, geometric, and process perspectives on facial cognition: Contexts and challenges. Lawrence Erlbaum Associates Inc., Hillsdale (1999)
2. Brunswik, E., Reiter, L.: Eindruckscharaktere Schematisierter Gesichter. Zeitschrift fur Psychologie 142, 67–135 (1937)
3. Sigala, N., Logothetis, N.K.: Visual Categorization Shapes Feature Selectivity in the Primate Temporal Cortex. Nature 415, 318–320 (2002)
4. Wilson, H.R., Loffler, G., Wilkinson, F.: Synthetic Faces, Face cubes, and the Geometry of Face Space. Vision Research 42, 2909–2923 (2002)
5. Bruce, V., Young, A.: In the Eye of the Beholder: the Science of Face Perception. Oxford University Press, Oxford (1998)
6. Tanaka, J.W., Farah, M.J.: Parts and Wholes in Face Recognition. Quarterly Journal of Experimental Psychology Section A—Human Experimental Psychology 46, 225–245 (1993)

7. Collishaw, S.M., Hole, G.J.: Featural and Configurational Processes in the Recognition of Faces of Different Familiarity. Perception 29, 893–909 (2000)
8. Tanaka, J.W., Sengco, J.A.: Features and Their Configuration in Face Recognition. Memory and Cognition 25, 583–592 (1997)
9. Costen, N.P., Parker, D.M., Craw, I.: Effects of High-pass and Low-pass Spatial Filtering on Face Identification. Perception and Psychophysics 58, 602–612 (1996)
10. Fiorentini, A., Maffei, L., Sandini, G.: The Role of High Spatial Frequencies in Face Perception. Perception 12, 195–201 (1993)
11. Gold, J., Bennett, P.J., Sekuler, A.B.: Identification of Bandpass Filtered Letters and Faces by Human and Ideal Observers. Vision Research 39, 3537–3560 (1999)
12. Hayes, T., Morrone, M.C., Burr, D.C.: Recognition of Positive and Negative Bandpass Filtered Images. Perception 15, 595–602 (1986)
13. Nasanen, R.: Spatial Frequency Bandwidth Used in the Recognition of Facial Images. Vision Research 39, 3824–3833 (1999)

Epipolar Geometry for Humanoid Robotic Heads

Justin Hart, Brian Scassellati, and Steven W. Zucker

Department of Computer Science
Yale University
New Haven, Connecticut 06520, USA
justin.hart@yale.edu, scaz@cs.yale.edu, steven.zucker@yale.edu

Abstract. Stereoscopic vision is a capability that supports the ability of robots to interact with visually complex environments. Epipolar geometry captures the projective relationship between the cameras in a stereo vision system, assisting in the reconstruction of three-dimensional information. However, a basic problem arises for robots with active vision systems whose cameras move with respect to each other: the epipolar geometry changes with this motion. Such problems are especially noticeable in work with humanoid robots, whose cameras move in order to emulate human gaze behavior. We develop an epipolar kinematic model that solves this problem by building a kinematic model based on the optical properties of a stereo vision system. We show how such a model can be used in order to update the epipolar geometry for the head of a humanoid robot.

1 Introduction

While stereo vision provides one of the richest feedback pathways for inferring such structure from our physical environment, to utilize advanced stereo computer vision techniques that are most relevant to biological perception [1] [2] requires knowledge of the imaging system's epipolar geometry. However the world rarely stands still, and on platforms where the cameras can move independently of one other the epipolar geometry will change with this motion. Here we develop epipolar kinematic models, or kinematic models that track the motion of optical properties of the system. The result is that motor data is used to compute an updated representation of the epipolar geometry. Particular emphasis in this paper is placed on computing such models for humanoid robotic heads.

Camera calibration is the process of measuring the parameters necessary for quantitative interaction with the 3D Euclidean world. The intrinsic parameters, which include focal length, principal point, and a skew factor relating the x and y axes, describe the camera itself. The extrinsic parameters, position and orientation, describe its pose in space. Additionally, lens distortion is often modeled. It has been a heavily researched topic in the computer vision and photogrammetry communities. [3] and [4] both provide excellent overviews of prior work and are seminal papers on the topic.

B. Caputo and M. Vincze (Eds.): ICVW 2008, LNCS 5329, pp. 24–36, 2008.

In a stereo vision system, epipolar geometry describes the projective relationship between two camera views, and can either be computed from their calibration [5], or estimated for uncalibrated cameras via methods such as the 8-point algorithm [6]. In the case of calibrated cameras, the epipolar geometry is described by the essential matrix. In the case of uncalibrated cameras, it is referred to as the funamental matrix.

An active vision system is a vision system in which either the cameras are able to move or they are attached to a device that is able to manipulate its environment. Such systems include cameras mounted on robotic arms, often referred to as *hand cameras*, and also in the heads of humanoid robots, such as our upper-torso humanoid infant, Nico, which is discussed in more depth in Section 5.1. The desire to calibrate the position of the cameras relative to the underlying robotic platform has given rise to two tasks, hand-eye and head-eye calibration, which either describe solving for how a camera is mounted with respect to a movable platform, usually with known kinematics, or solving for its position in space with respect to a manipulator, [7][8][9][10][11]. Kinematic calibration is the process of estimating the kinematics of the underlying system, [12][13].

Moving cameras present a unique challenge to robotics and vision researchers who wish to exploit the epipolar geometry of multiple cameras to perform stereo vision tasks. Such a scenario arises whenever a humanoid robot performs a saccade, a tracking motion, or when the eyes verge upon an attended to object. In this paper, we discuss the relationship between camera calibration, the estimation of epipolar geometry, and the kinematics of active vision systems. Prior work on this problem has focused on the use of 3D data to estimate ego-motion visually [14], tracking points in a stereo pair [15], or developing kinematic models by detaching the cameras from the head and viewing it using an external vision system [12][13].

The central contribution of this paper is the notion of an epipolar kinematic model, which is a kinematic model based on the motion of optical properties of the projective relationship between the cameras in a stereo active vision system as the cameras move through space. From this model we can compute current epipolar geometry using only knowledge of the current angles of the motors. To demonstrate, we will build such a model for our upper-torso humanoid. This model will be suitable for use with many. We present results from a preliminary implementation of the algorithm.

2   Background

2.1   The Pinhole Camera Model

Following standard notation, as found in [5], let X denote the homogeneous representation of a point in 3-space, and x its image. When discussing a stereo pair of cameras, determine one of the two cameras to be the *first* camera. All properties of the second camera will be marked with a $'$. For instance, let x

represent the image of X in the first camera and x' the image in the second camera.

The camera projection matrix, Equation 1, represents the projection of a 3D point, X, by a camera to a 2D point, x.

$$x = PX \tag{1}$$

Modeling the camera under the standard pinhole camera model, the camera calibration matrix, Equation 2 captures the camera's intrinsic parameters, which are properties of the camera itself. α and β, express focal length and are generally equal, and γ, which is the skew factor between the x and y axes, is generally 0. u_0 and v_0 represent the principal point. Together, they define the camera calibration matrix

$$A = \begin{bmatrix} \alpha & \gamma & u_0 \\ 0 & \beta & v_0 \\ 0 & 0 & 1 \end{bmatrix} \tag{2}$$

The extrinsic parameters, R, the rotation of the camera, and C, the camera center are combined with the camera calibration matrix as in Equation 3 to yield the camera projection matrix. These parameters can be retrieved via a number of standard camera calibration methods[4][3].

$$P = A[R \mid -RC] \tag{3}$$

2.2 Epipolar Geometry

Under the pinhole camera model, image points are represented as rays of light intersecting the *image plane* on a line running through the camera center. Given a pair of cameras, P and P', and a point x in camera P, we can constrain the position of x', the image of the same three-dimensional point X in P' to a line, l'. The image of one camera's camera center in the other camera is called an epipole. This system is called epipolar geometry, because these epipolar lines must all run through the epipole.

This relationship can be captured by the *fundamental matrix*, F, Equation 4.

$$x'^T F x = 0 \tag{4}$$

Given calibrated cameras, we can express our points as normalized image coordinates, coordinates corresponding to the same camera, but with A equal to the identity matrix. We express our coordinate system with in terms of P, giving us $P = [I|0]$ and $P' = [R|-RC]$. In this case, our essential matrix can be expressed as in Equation 5. The relationship between E and F is Equation 6. The fundamental and essential matrices can be computed using standard techniques [6][5][16][17].

$$E = [-RC]_\times R \tag{5}$$

$$F = A'^{-T} E A^{-1} \tag{6}$$

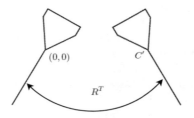

Fig. 1. Setup of the extrinsic parameters in the epipolar geometry problem. We define our coordinate system with the first camera at the origin. The second camera is rotated by R^T.

3   Epipolar Kinematics

In the case of stereo vision systems in which the cameras remain stationary with respect to each other, it is enough to estimate the epipolar geometry once, via the familiar process of matching control points in each image to each other and computing the projective relationship between them for 8 or more points [5]. For vision systems such as Nico's, however, this estimate will become inaccurate the first time that the robot moves its eyes.

Approaches have been demonstrated that cope with this via purely visual means, such as optical flow [15][14]. While a number of stereo tasks can still be performed via algorithms that do not require knowledge of epipolar geometry, such approaches ignore kinematic information available to the system that can be used to maintain the estimate in "real time." Most significantly, there is evidence that primate visual processing is structured precisely to take advantage of this [18].

From the formulation in Section 2.1, we can update the essential matrix with respect to camera motion provided that we know the way that the cameras move with respect to each other. Note that here we specifically mean the motion of the camera center and orientation of our pinhole cameras, optical properties that can be retrieved via standard computer vision techniques. One of the central insights of this work is that we can estimate our kinematic models based on these optical properties. This allows us to build our models using only image data processed by our stereo vision system with its joints turned in several orientations, rather than requiring for us to preprogram the system's kinematics or externally calibrate the kinematics of our visual system [12][13].

4   Epipolar Kinematics for a Humanoid

We define this model as a kinematic model over two revolute joints. This is reflective of those degrees of freedom relevant to the epipolar geometry of the head of our our humanoid robot, Nico, as well as those of many other humanoid robots.

Backlash is not modeled, and should be considered on top of this model if it is a significant concern. Finally, assume that the camera faces directly away from the axis of rotation of the joint on which it is mounted. We can easily eliminate this assumption, but retain it because it reduces the number of measurements that must be made, resulting in a faster calibration process, and also because it accurately describes the vision systems on most humanoid robots. We feel that the community of researchers working with humanoid robots is the most likely group to incorporate this method into their work. In Section 7, we will briefly discuss how to eliminate this assumption as well as how to model more complicated kinematic systems.

Our epipolar kinematic calibration algorithm is agnostic to the methods used for camera calibration and estimation of epipolar geometry. As such, we present this as a framework into which preferred methods for these two processes can be plugged in. By turning the linkage on which the camera is mounted and observing the relationship of this view to the view before turning the camera we can deduce the kinematics of the system. If that system has the constraint that the camera faces directly away from the center of rotation, as it does on Nico, then we are able to uncover the kinematics of that linkage by observing as few as two views.

4.1   Calibration Algorithm

Initial measurement. Proceed by choosing two angles for each of the eye motors to be calibrated. Denote the first camera in the first orientation, $Cam_{1,1}$, in the second orientation, $Cam_{1,2}$, the second camera in the first orientation

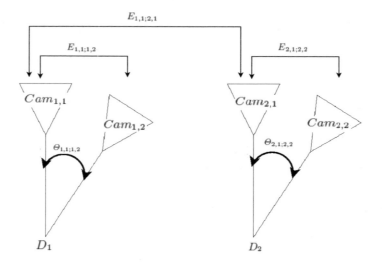

Fig. 2. Camera orientations and variables used in this process

$Cam_{2,1}$, and so forth for all parameters of the system, as in Figure 2. Let $E_{1,1;1,2}$ be the essential matrix between the $Cam_{1,1}$ and $Cam_{1,2}$, $E_{2,1;1,2}$ between $Cam_{1,1}$ and $Cam_{2,1}$ and so forth. As is standard, in the below discussion we will treat the first orientation in each essential matrix having its rotation matrix equal to the identity matrix and its camera center at the origin.

Calibrate the cameras and solve for $E_{1,1;1,2}$, $E_{2,1;2,2}$, and $E_{1,1;2,1}$.[1] Use any preferred method for both processes.[2]

Kinematic rotation axis and angle. Let $R_{1,1;1,2}$ be the rotation matrix found by decomposing $E_{1,1;1,2}$. If $V_{1,1;1,2}$ is the corresponding rotation vector, found by Rodrigues' Rotation Formula [19], then $\Theta_{1,1;1,2}$ is the magnitude of $V_{1,1;1,2}$, Equation 7.

$$\Theta = ||V|| \tag{7}$$

Dividing by Θ yields a unit axis of rotation, S, Equation 8.

$$S = \frac{V}{\Theta} = \frac{V}{||V||} \tag{8}$$

Kinematic link length. The camera centers for a camera before and after motion, such as $C_{1,1}$ and $C_{1,2}$, and the center of rotation of the epipolar kinematic linkage form an isosceles triangle. Therefore, the length of the linkage is given by Equation 9.

$$L_1 = \frac{\frac{||C_{1,2}||}{2}}{sin(\frac{\Theta_{1,1;1,2}}{2})} \tag{9}$$

Finding the center of rotation. Per our assumption that the camera faces directly away from the axis of rotation, compute the center of rotation via Equation 10, where D_1 is the center of rotation for for the first camera. D_2 is computed analogously.

$$D_1 = C_{1,1} - [0\ 0\ L_1] \tag{10}$$

Algorithm. Our entire calibration algorithm is summarized in Algorithm 1. This algorithm need be performed only once, at startup. The update rule to estimate the updated essential matrix is presented in Section 4.2.

---

[1] The purpose of the essential matrices is to ground the coordinate system with respect to the first camera orientation in each. If the calibration method places all of the extrinsic parameters in the same coordinate system, this step can be ignored, and the epipolar kinematic calibration process modified accordingly.

[2] The reader who is familiar with these processes will note that this might yield multiple values for A for the same physical camera, hopefully remarkably close to one another. In our implementation, we calibrate the cameras once using OpenCV.

Algorithm 1. Estimation of system parameters

---
1. Calibrate the two cameras in the stereo active vision system, Cam_1 and Cam_2
2. Choose 4 camera orientations, two for each camera, $Cam_{1,1}$, $Cam_{1,2}$, $Cam_{2,1}$, $Cam_{2,2}$
3. Estimate the essential matrix in each orientation
4. Factor $E_{1,1;1,2}$, $E_{2,1;2,2}$, $E_{1,1;2,1}$ per Equation 5
5. Compute $V_{1,1;1,2}$ and $V_{2,1;2,2}$ from $R_{1,1;1,2}$, $R_{2,1;2,2}$ via Rodrigues' Rotation Formula
6. Compute $\Theta_{1,1;1,2}$ and $\Theta_{2,1;2,2}$, per Equation 7
7. Compute S_1 and S_2 per Equation 8
8. Compute L_1 and L_2 per Equation 9
9. Compute D_1, D_2 per Equation 10

4.2   Updating the Essential Matrix

At runtime, we update our essential matrix to reflect the new position and orientation of the cameras each time they move with respect to each other. This means updating every time the motors move, changing this relationship.

Let $\Theta_{1,Enc}$ be the difference between the Θ indicated by the encoder at $Cam_{1,1}$ and the current encoder reading for that camera's associated motor. Let $\Theta_{2,Enc}$ be the analogous value for the second camera. All variables subscripted Enc will be with respect to the current encoder reading. Compute updated $V_{1,Enc}$, $V_{2,Enc}$ by Equation 11.

$$V_{1,Enc} = \Theta_{1,Enc} * S_1 \tag{11}$$

Update $R_{1,Enc}$ and $R_{2,Enc}$ via Rodrigues' Rotation Formula. We'll denote variables reflective of the current position and rotation of the second camera with respect to the current position and rotation of the first camera by subscripting them $CurSys$. Let R_{CurSys} is given by Equation 12.

$$R_{CurSys} = R_{1,Enc}^T * R_{2,Enc} * R_{1,1;2,1} \tag{12}$$

Find the updated camera centers, $C_{1,Enc}$ and $C_{2,Enc}$ via Equation 13.

$$C_{1,Enc} = R_{1,Enc}^T [0\ 0\ L_1] - D_1 \tag{13}$$

Find the updated camera center in the second view with respect to the first view, C_{CurSys}, Equation 14. Remember that $C_{1,Enc}$ and $C_{2,Enc}$ do not share the same world coordinate system.

$$C_{CurSys} = C_{2,1} - C_{1,Enc} + C_{2,Enc} \tag{14}$$

Compute the updated essential matrix, Equation 15.

$$E_{CurSys} = A_2'^{-T} [-R_{CurSys} * C_{CurSys}]_\times R_{CurSys} A_1^{-1} \tag{15}$$

Our entire update algorithm is summarized in Algorithm 2. Since this algorithm involves only constant-time matrix computations, it can be used to update the epipolar geometry in real-time.

Algorithm 2. Essential matrix update

1. Compute $V_{1,Enc}$, $V_{2,Enc}$ per Equation 11. Compute updated $R_{1,Enc}$, $R_{2,Enc}$ from $V_{1,Enc}$, $V_{2,Enc}$ via Rodrigues' Rotation Formula
2. Compute R_{CurSys} via Equation 12
3. Compute $C_{1,Enc}$, $C_{2,Enc}$ via Equation 13
4. Compute C_{CurSys} via Equation 14
5. Compute E_{CurSys} via Equation 15

5 Tests

5.1 Platform

Nico, Figure 3, is an upper-torso humanoid robot that has been modeled after the the kinematic structure of a fiftieth percentile 12-month-old male infant. It has 23 mechanical degrees of freedom, including six in each arm and two in its recently-added hand. Its head has six degrees of freedom and employs a foveated vision system consisting of four NTSC color cameras mounted in two anthropomorphic eyes. The eyes have mechanically coupled pitch and independent yaw degrees of freedom. Nico's compute platform includes a 20-node cluster running the QNX real-time operating system. Nodes are connected via 100 Mbit Ethernet to each other and to a number of Linux and Windows machines whose configurations change from experiment to experiment. This architecture allows us to easily

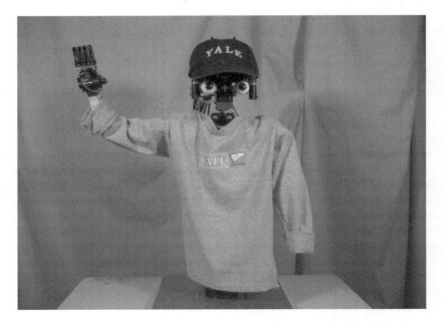

Fig. 3. Nico, an upper-torso humanoid infant

integrate software packages on multiple platforms into Nico's control system, as well as to remotely operate Nico via the Internet.

5.2   Test Setup

In order to test our system we took 3 sets of images of chessboards. Imaging proceeded as follows. The cameras were first to -10 degrees, then to 0 degrees, 5 degrees, and 10 degrees. A chessboard was placed in front of the robot and it was visually confirmed that the robot could locate all of the interior corners of the chessboard in both cameras in the last 3 of these orientations. The first orientation was then returned to in order to assure that any backlash was worked out of the system. It is expected that this, combined with the fact that the system uses zero-backlash motors worked out most of the backlash. Several images were taken in each position for each orientation of the chessboard in order to generate a set of images for both camera calibration and the estimation of epipolar geometry.

Using these images, the cameras were calibrated using OpenCV [20], and the essential matrices between the 0 and 10 degree views for each motor, and the 0 degree views for both motors were computed using a third-party tool, *Gandalf*.[3]

Tests were performed on the images shot at 5 degrees. For comparison, we computed the essential matrix both using the epipolar kinematics algorithms and directly from the images.

6   Results

The results included in this section should be regarded as preliminary, as the accuracy in our essential matrix estimates does not match the sub-pixel resolution expected from state of the art algorithms. Updated results will be made available in a future publication.

Upon testing, we found that the software package we used to estimate the essential matrix, Gandalf, exhibits a degree of numerical instability that is quite common in software that is used to estimate epipolar geometry [21]. In order to work around this instability, we built a software package that processed all possible subsets of image pairs for each essential matrix to Gandalf. As an error metric, we adopted the mean distance in pixels between an epipolar line and its corresponding point. We computed the epipolar lines in the right image corresponding to chessboard corners in the left image and measured the distance to the corresponding image point. We chose each essential matrix as the one corresponding to the lowest mean distance for each matrix required to compute the epipolar kinematic model, as well as for the essential matrix computed directly from the test images. See Table 2 for the mean distance corresponding to each matrix.

The redundancy of the Θ yielded by both checking the encoder readings and the computation of the essential matrix gives us the opportunity to check our robot's physical readings against those estimated by the vision algorithm. Results of this comparison are listed in Table 1. As we can see, there is significant

---

[3] http://gandalf-library.sourceforge.net/

(a) Left camera image.

(b) Right camera image, with epipolar lines estimated directly from the images.

(c) Right camera image, with epipolar lines estimated using the essential matrix update algorithm.

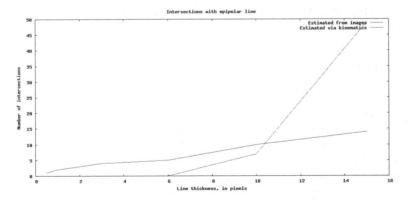

Fig. 4. Number of intersections from epipolar line to corresponding point plotted against line thickness in pixels. There are 49 test points.

disagreement between the vision algorithm and the motors. Potential sources of this error include backlash in the motors and gears, and error in the estimates of the essential matrix or camera calibration.

Table 1. Θ's estimated by essential matrix computation

	Θ_a	Θ_b	Θ_c
Estimated from E	0.17412	0.0479997	0.0109251
Turned to	-	10	10

Table 2. Mean distance from a given epipolar line to its corresponding point

Essential Matrix	Mean distance from point to epipolar line
Input to the epipolar kinematic model	
$E_{1,1;1,2}$	1.11736
$E_{2,1;2,2}$	0.638143
$E_{1,1;2,1}$	2.38185
Results	
Directly Computed From Images	18.7897
Epipolar Kinematic Model	11.0749

The epipolar lines yielded by these computations appear in Figures 4(a), 4(b), and 4(c). As we can see, the positioning of the epipole in the two images is not the same. As a test of the relative quality of the two algorithms, we took the mean distance from the epipolar line computed from the chessboard corners in the first image to the corresponding point in the second image using both an essential matrix estimated directly from imaged points and one estimated using an epipolar kinematic model. Results can be seen in Table 2. Another rough estimate of the quality of the essential matrix is the number of epipolar lines that intersect with their corresponding image points. We compare the two matrices in Figure 4.

7   Conclusion

The primary insight offered in this paper is that we can build epipolar kinematic systems, kinematic systems built directly off of the optical properties of a stereo vision system, in order to track these properties as the system moves. This allows us to keep a consistent view of the epipolar geometry of the system as it undergoes motion.

To demonstrate this technique, we showed how to compute the epipolar kinematics of the degrees of freedom of the active vision head on our humanoid robot, Nico, for those degrees of freedom effecting the system's epipolar geometry. The algorithms in this paper are suitable for the active vision heads of many humanoid robots. In this exploration we can clearly see that the estimation of epipolar kinematics is built on top of the existing suite of techniques available to the tasks of camera calibration and estimation of epipolar geometry.

Though a version of this algorithm that uses two orientations per camera is presented in this paper, it is possible to update this algorithm to use three orientations per camera in order to lift the assumption that the cameras face straight forward from center of rotation. In this case, we are able to estimate the circle defining the rotation from the three camera centers, found during the estimation of epipolar geometry, per camera, leaving us only to solve for the rotation of the camera with respect to the endpoint of the linkage. Such an algorithm is equivalent to solving head/hand-eye calibration and kinematic calibration simultaneously. Building on this process, we can estimate the epipolar kinematics of systems where more than one linkage can control the orientation of the camera. The same mathematics can equivalently be used to solve for the kinematics of a manipulator, or other kinematic linkage visible in the visual field, and its relationship to the coordinate system of the visual system. This is all deferred to future work in which the robot learns about its self in terms of its kinematics and its sensors.

Acknowledgements

Support for this work was provided by a National Science Foundation CAREER award (#0238334) and NSF award #0534610 (Quantitative Measures of Social Response in Autism). This research was supported in part by a software grant from QNX Software Systems Ltd and support from the Sloan Foundation. S.W.Z. is additionally supported by AFOSR, NSA, and NSF.

References

1. Li, G., Zucker, S.W.: Contextual inference in contour-based stereo correspondence. Int. J. of Computer Vision 69(1), 59–75 (2006)
2. Li, G., Zucker, S.W.: Differential geometric consistency extends stereo to curved surfaces. In: Proceedings of the 9th European Conference on Computer Vision, pp. III: 44–57 (2006)
3. Zhang, Z.: A flexible new technique for camera calibration. IEEE Transactions on Pattern Analysis and Machine Intelligence 22(11), 1330–1334 (2000)
4. Tsai, R.Y.: A versatile camera calibration technique for high-accuracy 3d machine vision metrology using off-the-shelf tv cameras and lenses. IEEE Journal of Robotics and Automation, 221–244 (1992)
5. Hartley, R.I., Zisserman, A.: Multiple View Geometry in Computer Vision, 2nd edn. Cambridge University Press, Cambridge (2004)
6. Deriche, R., Zhang, Z., Luong, Q.T., Faugeras, O.D.: Robust recovery of the epipolar geometry for an uncalibrated stereo rig. In: Eklundh, J.-O. (ed.) ECCV 1994. LNCS, vol. 800, pp. 567–576. Springer, Heidelberg (1994)
7. Shih, S.W., Hung, Y.P., Lin, W.S.: Head/eye calibration of a binocular head by use of single calibration point. In: Proceedings of the IEEE Southwest Symposium on Image Analysis and Interpretation, pp. 154–159 (1994)
8. Li, M., Betsis, D.: Head-eye calibration. In: ICCV 1995: Proceedings of the Fifth International Conference on Computer Vision, Washington, DC, USA, p. 40. IEEE Computer Society, Los Alamitos (1995)

9. Tsai, R., Lenz, R.: Real time versatile robotics hand/eye calibration using 3d machine vision. In: Proceedings of IEEE International Conference on Robotics and Automation, 1998, April 24-29, 1988, vol. 1, pp. 554–561. IEEE Computer Society Press, Los Alamitos (1988)
10. Shiu, Y., Ahmad, S.: Calibration of wrist-mounted robotic sensors by solving homogeneous transform equations of the form ax=xb. IEEE Transactions on Robotics and Automation 5(1), 16–29 (1989)
11. Tsai, R., Lenz, R.: A new technique for fully autonomous and efficient 3d robotics hand/eye calibration. IEEE Transactions on Robotics and Automation 5(3), 345–358 (1989)
12. Shih, S.W., Jin, J.S., Wei, K.H., Hung, Y.P.: Kinematic calibration of a binocular head using stereo vision with the complete and parametrically continuous model. In: Casasent, D.P. (ed.) Proc. SPIE, Intelligent Robots and Computer Vision XI: Algorithms, Techniques, and Active Vision, November 1992. The Society of Photo-Optical Instrumentation Engineers (SPIE) Conference, vol. 1825, pp. 643–657 (1992)
13. Shih, S.-W., Hung, Y.-P., Lin, W.-S.: Kinematic parameter identification of a binocular head using stereo measurements of single calibration point. Proceedings of the IEEE International Conference on Robotics and Automation 2, 1796–1801 (1995)
14. Bjorkman, M., Eklundh, J.-O.: A real-time system for epipolar geometry and egomotion estimation. In: Proceedings of the IEEE Conference on Computer Vision and Pattern Recognition, vol. 2, pp. 506–513 (2000)
15. Björkman, M., Eklundh, J.O.: Real-time epipolar geometry estimation of binocular stereo heads. IEEE Trans. Pattern Anal. Mach. Intell. 24(3), 425–432 (2002)
16. Luong, Q.-T., Faugeras, O.: The fundamental matrix: Theory, algorithms, and stability analysis. International Journal of Computer Vision 17(1), 43–75 (1996)
17. Hartley, R.I.: In defence of the 8-point algorithm. In: ICCV 1995: Proceedings of the Fifth International Conference on Computer Vision, Washington, DC, USA, p. 1064. IEEE Computer Society Press, Los Alamitos (1995)
18. Dobbins, A.C., Jeo, R.M., Fiser, J., Allman, J.M.: Distance modulation of neural activity in the visual cortex. Science (281), 552–555 (1998)
19. Faugeras, O.: Three-dimensional computer vision: a geometric viewpoint. MIT Press, Cambridge (1993)
20. Intel Corporation: Open Source Computer Vision Library: Reference Manual (1999-2001)
21. Izquierdo, E., Guerra, V.: Estimating the essential matrix by efficient linear techniques. IEEE Transactions on Circuits and Systems for Video Technology 13(9), 925–935 (2003)

Monitoring Activities of Daily Living (ADLs) of Elderly Based on 3D Key Human Postures

Nadia Zouba, Bernard Boulay, Francois Bremond, and Monique Thonnat

INRIA Sophia Antipolis, PULSAR Team, 2004, route des Lucioles, BP93, 06902
Sophia Antipolis Cedex, France

Abstract. This paper presents a cognitive vision approach to recognize
a set of interesting activities of daily living (ADLs) for elderly at home.
The proposed approach is composed of a video analysis component and
an activity recognition component.

A video analysis component contains person detection, person tracking
and human posture recognition. A human posture recognition is composed
of a set of postures models and a dedicated human posture recognition
algorithm.

Activity recognition component contains a set of video event models
and a dedicated video event recognition algorithm.

In this study, we collaborate with medical experts (gerontologists from
Nice hospital) to define and model a set of scenarios related to the interesting activities of elderly. Some of these activities require to detect
a fine description of human body such as postures. For this purpose, we
propose ten 3D key human postures usefull to recognize a set of interesting human activities regardless of the environment. Using these 3D key
human postures, we have modeled thirty four video events, simple ones
such as "a person is standing" and composite ones such as "a person is
feeling faint". We have also adapted a video event recognition algorithm
to detect in real time some activities of interest by adding posture.

The novelty of our approach is the proposed 3D key postures and the
set of activity models of elderly person living alone in her/his own home.

To validate our proposed models, we have performed a set of experiments in the Gerhome laboratory which is a realistic site reproducing
the environment of a typical apartment. For these experiments, we have
acquired and processed ten video sequences with one actor. The duration of each video sequence is about ten minutes and each video contains
about 4800 frames.

Keywords: 3D human posture, posture models, event models, ADLs.

1 Introduction

The elderly population is expected to grow dramatically over the next 20 years.
The number of people requiring care will grow accordingly, while the number
of people able to provide this care will decrease. Without receiving sufficient
care, elderly are at risk of loosing their independence. Thus a system permitting

B. Caputo and M. Vincze (Eds.): ICVW 2008, LNCS 5329, pp. 37–50, 2008.

elderly to live safely at home is more than needed. Medical professionals believe that one of the best ways to detect emerging physical and mental health problems, before it becomes critical - particularly for the elderly - is analyzing the human behavior and looking for changes in the activities of daily living (ADLs). Typical ADLs include sleeping, meal preparation, eating, housekeeping, bathing or showering, dressing, using the toilet, doing laundry, and managing medications. As a solution to this issue, we propose an approach which consists in the modeling of ten 3D key human postures useful to recognize some interesting activities of elderly. The recognition of these postures is based on using the human posture recognition algorithm proposed in [1] which can recognize in real time human postures with only one static camera regardless of its position.

In this paper, we focus on recognizing activities that elderly are able to do (e.g. ability of elderly person to reach and open a kitchen cupboard). The recognition of these interesting activities helps medical experts (gerontologists) to evaluate the degree of frailty of elderly by detecting changes in their behavior patterns. We also focus on detecting critical situations of elderly (e.g. feeling faint, falling down), which can indicate the presence of health disorders (physical and/or mental). The detection of these critical situations can enable early assistance of elderly.

In this paper, section 2 briefly reviews previous work on human posture recognition and activity recognition using video cameras. Section 3 describes our activity recognition approach. Results of the approach are reported in section 4. Finally, conclusion and future works are presented in section 5.

2   State of the Art

In this section we present firstly previous work on human posture recognition using 2D and 3D approaches, and secondly previous work on activity recognition.

2.1   Human Posture Recognition by Video Cameras

The vision techniques to determine human posture can be classified according to the type of model used (explicit, statistical, ...) and the dimensionality of the work space (2D or 3D). The **2D approaches** with **explicit models** [2] try to detect some body parts. They are sensitive to segmentation errors. The **2D approaches** with **statistical models** [3] are then proposed to handle the problems due to segmentation. These two 2D approaches are well adapted for real time processing but they depend on the camera view point. The **3D approaches** can also be classified in **statistical** and **model** based techniques. They consist in computing the parameters of the 3D model, such as the model projection on the image plane fits with the input image (often the silhouette). Some approaches compare the contour of the input silhouette with one of the projected model. In [4], the authors propose a method to reconstruct human posture from un-calibrated monocular image sequences. The human body articulations are extracted and annotated manually on the first image of a video sequence, then image processing techniques (such as linear prediction or least square matching) are used to extract articulations from the other frames. The learning-based approaches avoid the need of an explicit 3D human body model. In [5], the authors propose a learning-based method for recovering 3D human body posture from single images and monocular image sequences. These 3D approaches are partially independent from the camera view point but they need to define many parameters to model the human posture.

In this study, we choose to use an **hybrid approach** described in [1]. This approach combines the advantages of the 2D and 3D approaches to recognize the entire human body postures in real-time. It is based on a 3D human model and is independent from the point of view of the camera and employs silhouette represented from 2D approaches to provide a real-time processing.

2.2 Activity Recognition

Previous activity detection research focused on analyzing individual human behaviors. **Rule-based methods** proposed in [6] have shown their merits in action analysis. Rule-based systems may have difficulties in defining precise rules for every behavior because some behaviors may consist of fuzzy concepts. **Statistical approaches**, from template models, linear models, to graphic models, have been used in human activity analysis. Yacoob and Black [7] used linear models to track cyclic human motion. Jebara and Pentland [8] employed conditional Expectation Maximization to model and predict actions. Aggarwal et al. [9] has reviewed different methods for human motion tracking and recognition. The probabilistic and stochastic approaches include HMM (Hidden Markov Model) and NNs (Neuronal Networks). They are represented by graphs. Hidden Markov models [10] have been used for recognizing actions and activities, and illustrated their advantages in modeling temporal relationships between visual-audio events. Chomat and Crowley [11] proposed a probabilistic method for recognizing activities from local spatio-temporal appearance. Intille and Bobick [12] interpret actions using Bayesian networks among multiple agents. Bayesian networks can combine uncertain temporal information and compute the likelihood for the trajectory of a set of objects to be a multi-agent action. Recently, Jesse Hoey et al. [13] successfully used only cameras to assist person with dementia during handwashing. The system uses only video inputs, and combines a Bayesian sequential estimation framework for tracking hands and towel, with a decision using a partially observable Markov decision process. Most of these methods mainly focus on a specific human activity and their description are not declarative and it is often difficult to understand how they work (especially for NNs). In consequence, it is relatively difficult to modify them or to add a priori knowledge. The **deterministic approaches** use a priori knowledge to model the events to recognize [14]. This knowledge usually corresponds to rules defined by experts from the application domain. These approaches are easy to understand but their expressiveness is limited, due to the fact that the variety of the real world is difficult to represent by logic. The **approaches based on constraint resolution** are able to recognize complex events involving multiple actors having complex temporal relationships.

In this work we have used the approach described in [15]. This approach is based on constraint resolution. It uses a declarative representation of events which are defined as a set of spatio-temporal and logical constraints. This technique is easy to understand since it is based on constraints which are defined in a declarative way.

The next section presents the approach for activity recognition we used in this paper.

3   The Proposed Activity Recognition Approach

3.1   Overview

The proposed approach is composed of (1) a video analysis component which contains person detection, person tracking and a human posture recognition, (2) an activity recognition component which contains a set of video event models and a video event recognition algorithm. A simplified scheme of the proposed approach is given in figure 1. Firstly, we present the video analysis component and secondly the activity recognition component.

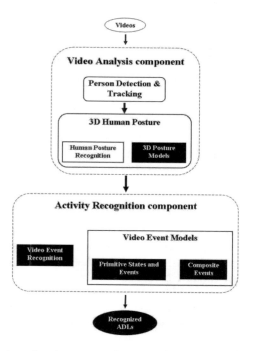

Fig. 1. The architecture for the proposed approach. The contribution of this paper is represented with black background.

3.2   Video Analysis

In this section we describe shortly person detection and tracking method. We also describe the used human posture recognition algorithm and we detail the proposed 3D posture models.

Person Detection and Tracking. For detecting and tracking person we use a set of vision algorithms coming from a video interpretation platform described in [16]. A first algorithm segments moving pixels in the video into a binary image by subtracting the current image with the reference image. The reference image is updated along the time to take into account changes in the scene (light, object displacement, shadows). The moving pixels are then grouped into connected regions, called blobs. A set of 3D features such as 3D position, width and height are computed for each blob. Then the blobs are classified into predefined classes (e.g. person). After that the tracking task associates to each new classified blob a unique identifier and maintains it globally throughout the whole video. Figure 2 illustrates the detection, classification and tracking of a person in the experimental laboratory.

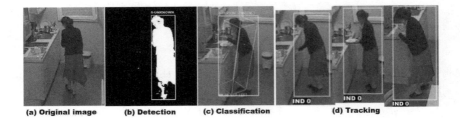

(a) Original image   (b) Detection   (c) Classification   (d) Tracking

Fig. 2. Detection, classification and tracking of a person. (a) represents the original image. (b) the moving pixels are highlighted in white and clustered into a mobile object enclosed in an orange bounding box. (c) the mobile object is classified as a person. (d) shows the individual identifier (IND 0) and a colored box associated to the tracked person.

3D Human Posture Recognition. In this section, we firstly present the human posture recognition algorithm and secondly the proposed 3D posture models.

- **Human Posture Recognition Algorithm:** We have used a human posture recognition algorithm [1] in order to recognize in real time a set of human postures once the person evolving in the scene is correctly detected. This algorithm determines the posture of the detected person using the detected silhouette and its 3D position. The human posture recognition algorithm is based on the combination between a set of 3D human model with a 2D approach. These 3D models are projected in a virtual scene observed by a virtual camera which has the same characteristics (position, orientation and

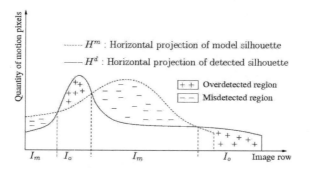

Fig. 3. Horizontal projection of model and detected silhouettes. I_o (resp. I_m) represents the overdetected (resp. misdetected) region.

field of view) than the real camera. The 3D human silhouettes are then extracted and compared with the detected silhouette using a 2D techniques (projection of the silhouette pixels on the horizontal and vertical axes, see figure 3).

$$R_o(H) = \frac{\sum_{i \in I_o}(H_i^d - H_i^a)^2}{\sum_i (H_i^d)^2} \ , \ R_m(H) = \frac{\sum_{i \in I_m}(H_i^d - H_i^a)^2}{\sum_i (H_i^a)^2} \qquad (1)$$

$$(\alpha_1, \alpha_2, \beta_1, \beta_2) \in [0,1]^4, \alpha_1 + \alpha_2 + \beta_1 + \beta_2 = 1$$

The distance from the detected silhouette to the model silhouette is computed with the equation (2):

$$dist(S_a, S_d) = \alpha_1 R_o(H) + \beta_1 R_m(H) + \alpha_2 R_o(V) + \beta_2 R_m(V) \qquad (2)$$

The most similar extracted 3D silhouette corresponds to the current posture of the observed person. This algorithm is real time (about eight frames per second), requires only a fix video camera and do not depend on the camera position.

– **3D Posture Models:** The **posture models** are based on a 3D geometrical human model. We propose ten 3D key human postures which are useful to recognize activities of interest. These postures are displayed in figure 4: standing (a), standing with arm up (b), standing with hands up (c), bending (d), sitting on a chair (e), sitting on the floor with outstretched legs (f), sitting on the floor with flexed legs (g), slumping (h), lying on the side with flexed legs (i), and lying on the back with outstretched legs (j). Each of these postures plays a significant role in the recognition of the targeted activities of daily living. For example, the posture "standing with arm up" is used to detect when a person reaches and opens kitchen cupboard and her/his ability to do it. The posture "standing with hands up" is used to detect when a person is carrying an object such as plates. These proposed human postures are not an exhaustive list but represent the key human postures taking part in everyday activities.

Fig. 4. The proposed 3D human postures

3.3   Activity Recognition

In this section, we firstly describe the video event recognition algorithm and secondly we present the proposed video event models.

Video Event Recognition Algorithm. The video event recognition algorithm detects which event is happening from a stream of observed persons tracked by a vision component at each instant. The recognition process takes as input the a priori knowledge of the scene and the event models.

An event is composed of actors, sub-events and constraints. An actor can be a person tracked as a mobile object by the vision component or a static object of the observed environment like a chair. A person is represented by her/his characteristics: her/his position in the observed environment, width, velocity,... A static object of the environment is defined by a priori knowledge and can be either a zone of interest (e.g. the entrance zone) or a piece of equipment (a 3D object such as a table). A zone is represented by its vertices and a piece of equipment is represented by a 3D bounding box. The zones and the equipment constitute the scene context of the observed environment.

To recognize the pre-defined events at each instant, the algorithm verifies that the sub-events are recognized and the constraints are satisfied.

The video event recognition algorithm is based on the method described in [15]. We have adapted this algorithm to detect in real time some activities of interest by adding posture.

Video Event Models. The **event models** are defined using an event description language designed in a generic framework [16]. The video event model corresponds to the modeling of all the knowledge used by the system to detect video events occurring in the scene. The description of this knowledge is declarative and intuitive (in natural terms), so that the experts of the application domain can easily define and modify it. Four types of video events (called **components**) can be defined: primitive states, composite states, primitive events and composite events. A state describes a stable situation in time characterizing one or several physical objects (i.e. actors). A **primitive state** (e.g. a person is located inside a zone) corresponds to a perceptual property directly computed by the vision components. A **composite state** is a combination of primitive states. An event is an activity containing at least a change of state values between two consecutive times. A **primitive event** corresponds to a change of primitive state values (e.g. a person changes a zone). A **composite event** is a combination of primitive states and/or primitive events (e.g. preparing meal). As general model, a video event model is composed of five parts: "**physical objects**" involved in the event (e.g. person, equipment, zones of interest), "**components**" corresponding to the sub-events composing the event, "**forbidden components**" corresponding to the events which should not occur during the main event, "**constraints**" are conditions between the physical objects and/or the components (including symbolic, logical, spatial and temporal constraints including Allens interval algebra operators [17]), and "**alarms**" describe the actions to be taken when the event is recognized.

In the framework of homecare monitoring, in collaboration with gerontologists, we have modeled several primitive states, primitive events and composite events. First we are interesting in modeling event characteristic of critical situations such as falling down. Second, these events aim at detecting abnormal changes of behavior patterns such as depression. Given these objectives we have selected the activities that can be detected using video cameras. For instance, the detection of **"gas stove on"** when a person is doing a different activity for a long time is interesting but cannot be easily detected by only video cameras and requires additional information such as the one provided by environmental sensors (e.g. gas consumption sensor). In this paper we are focusing only on video cameras and contributions with other environmental sensors for activity recognition belong to other ongoing work.

In this work, we have modeled **thirty four video events**. In particular, we have defined fourteen primitives states, four of them are related to the location of the person in the scene (e.g. inside kitchen, inside livingroom) and the ten remaining are related to the proposed 3D key human postures. We have defined also four primitive events related to the combination of these primitive states: **"standing up"** which represents a change state from sitting or slumping to standing, **"sitting down"** which represents a change state from standing, or bending to sitting on a chair, **"sitting up"** represents a change state from lying to sitting on the floor, and **"lying down"** which represents a change state from standing or sitting on the floor to lying. We have defined also six primitive events such as: stay in kitchen, stay in livingroom. These primitive states and events are used to define more composite events.

For this study, we have modeled ten composite events. In this paper, we present just two of them: **"feeling faint"** and **"falling down"**.

There are different visual definition for describing a person falling down. Thus, we have modeled the event "falling down" with three models:

Falling down 1: A change state from standing, bending, sitting on the floor(with flexed or outstretched legs) and lying (with flexed or outstretched legs).
Falling down 2: A change state from standing, and lying (with flexed or outstretched legs).
Falling down 3: A change state from standing, bending and lying (with flexed or outstretched legs).

The model of the "feeling faint" event is shown bellow. The "feeling faint" model contains three 3D human postures components, involves one person and additional constraints between these components.

```
CompositeEvent(PersonFeelingFaint,
PhysicalObjects( (p: Person) )
Components( (pStand: PrimitiveState Standing(p))
(pBend: PrimitiveState Bending(p))
(pSit: PrimitiveState Sitting_Outstretched_Legs(p)) )
Constraints( (pStand; pBend; pSit)
(pSit's Duration >=10))
Alarm(AText(''Person is Feeling Faint")
AType(''URGENT")) )
```

"Feeling faint" model

The following text shows an example of the definition of the model "falling down 1".

```
CompositeEvent(PersonFallingDown1,
PhysicalObjects( (p: Person) )
Components( (pStand: PrimitiveState Standing(p))
(pBend: PrimitiveState Bending(p))
(pSit: PrimitiveState Sitting_Flexed_Legs(p))
(pLay: PrimitiveState Lying_Outstretched_Legs(p)) )
Constraints( (pSit before_meet p_Lay)
(pLay's Duration >=50))
Alarm(AText(''Person is Falling Down")
AType(''VERYURGENT")) )
```

"Falling down 1" model

In this approach we have proposed ten 3D key human postures and thirty four video event models useful to recognize a set of ADLs of elderly living alone in her/his own home. In the next section we present experiments we have done in the Gerhome laboratory and the obtained results.

4   Results and Evaluation

This section describes and discusses the experimental results. First, we describe the experimental site we have used to validate our approach and models. Then we show and discuss the results of activity recognition.

4.1 Experimental Site

Developing and testing the impact of the activity monitoring solutions requires
a realistic near-life environment in which training and evaluation can be performed. To attain this goal we have set up an experimental laboratory (Gerhome
laboratory) to analyze and evaluate our approach. This laboratory is located in
the CSTB (Centre Scientifique de Techniques du Batiment) at Sophia Antipolis. It looks like a typical apartment of an elderly person: $41m^2$ with entrance,
livingroom, bedroom, bathroom, and kitchen. The kitchen includes an electric
stove, microwave oven, fridge, cupboards, and drawers. 4 video cameras are installed in Gerhome laboratry. One video camera is installed in the kitchen, two
video cameras are installed in the livingroom and the last one is installed in
the bedroom to detect and track the person in the apartment and to recognize
her/his postures. This laboratory plays an important role in research and system
development in the domain of activity monitoring and of assisted living. Firstly,
it is used to collect data from the different installed video cameras. Secondly,
it is used as a demonstration platform in order to visualize the system results.
Finally, it is used to assess and test the usability of the system with elderly.
Currently, in this experiment, we have collected and processed data acquired by
one video camera. The 3D visualization of Gerhome laboratory is illustrated in
Figure 5.

4.2 Experimental Results

To validate our models, we have performed a set of human behaviors in the
Gerhome laboratory. For this experiment, we have acquired ten videos with one
human actor. The duration of each video is about ten minutes and each video
contains about 4800 frames (about eight frames per second).

For performance evaluation, we use classical metrics. When the system correctly claims that an activity occurs, a true positive (TP) is scored; a false
positive (FP) is scored when an incorrect activity is claimed. If an activity occurs and the system does not report it, a false negative (FN) is scored. We then
used the precision and sensitivity standard metrics to summarize the system
effectiveness. Precision is the ratio TP/(TP + FP), and sensitivity is the ratio
TP/(TP + FN).

The results of the recognition of the primitive states and events are presented
in table 1. The primitive states "in the kitchen" and "in the livingroom" are
well recognized by video cameras. The few errors in the recognition occur at
the border between livingroom and kitchen. These errors are due to noise and
shadow problems.

The preliminary results of the recognition of the different postures (a, b, c,
e, f, g, h, i, j) are encouraging. The errors in the recognition of these postures
occur when the system mixes the recognized postures (e.g. the bending posture
instead the sitting one). These errors are due to the segmentation errors (shadow,
light change, ...) and to object occlusions. To solve these errors, we plan to use
temporal filtering in the posture recognition process.

Table 1. Results for recognition of a set of primitive states and events

States and events	Ground truth	#TP	#FN	#FP	Precision	Sensitivity
In the kitchen	45	40	5	3	93%	88%
In the livingroom	35	32	3	5	86%	91%
Standing (a, b, c)	120	95	25	20	82%	79%
Sitting(e, f, g)	80	58	22	18	76%	72%
Slumping(h)	35	25	10	15	62%	71%
Lying (i, j)	6	4	2	2	66%	66%
Bending (d)	92	66	26	30	68%	71%
Standing up	57	36	21	6	85%	63%
Sitting down	65	41	24	8	83%	63%
Sitting up	6	4	2	1	80%	66%

Fig. 5. Recognition of two video events in Gerhome laboratory. (a) 3D visualization of the experimental site "Gerhome", (b) person is in the livingroom, (c) person is sitting in the floor with outstretched legs.

Fig. 6. Recognition of the "feeling faint" event. (a) person is standing. (b) person is bending. (c) sitting with outstretched legs.

We show in the figure 5 the recognition of the localization of the person inside livingroom and the recognition of the posture "sitting in the floor with outstretched legs". In the ten acquired videos, we have filmed one "falling down" event and two "feeling faint" events which have been correctly recognized.

Figure 6 and figure 7 show respectively the camera view and the 3D visualization of the recognition of the "feeling faint" event. Figure 8 and figure 9 show respectively the camera view and the 3D visualization of the recognition of the "falling down" event.

Fig. 7. 3D visualization of the recognition of the "feeling faint" event. (a) person is standing. (b) person is bending. (c) sitting with outstretched legs".

Fig. 8. Recognition of the "falling down" event. (a) person is standing. (b) sitting with flexed legs. (c) lying with outstretched legs.

Fig. 9. 3D visualization of the recognition of the "falling down" event. (a) person is standing. (b) sitting with flexed legs. (c) lying with outstretched legs.

5   Conclusions and Future Works

In this paper we have described a cognitive vision approach to recognize a set of activities of interest of elderly at home by using ten 3D key human postures. This approach takes as input only video data and produces as output the set of the recognized activities.

The first contribution of this work consists in the identifying and modeling of the ten 3D key human postures. These key postures are useful in the recognition of a set of normal and abnormal activities of elderly living alone at home. The second contribution of this work is the modeling of some activities of interest of elderly.

The proposed approach is currently experimented on small datasets, but we will next validate the performance of our approach on larger datasets (long videos on a long term period with different persons). Currently 14 different video sequences (more than 56 hours) have been recorded with elederly people in Gerhome laboratory. We plan two other different studies with elderly (+ 70 years) which will take place during the next six months. The first study will be performed in a clinical center (e.g. hospital, nursing home), and the second one in a free living environment (home of elderly people). We also plan to study some other postures and take account gestures in order to detect finer activities (e.g. kneeling).

Moreover, we envisage to analyze which sensors in addition to video cameras are the best for monitoring most activities of daily living. Our ultimate aim is to determine the best set of sensors according to various criteria such as cost, number of house occupants and presence of pets.

We also envisage to facilitate incorporation of new sensors by developing a generic model of intelligent sensor and to add the data uncertainty and imprecision on sensor measurement analysis.

References

1. Boulay, B., Bremond, F., Thonnat, M.: Applying 3d human model in a posture recognition system. Pattern Recognition Letter (2006)
2. Wren, C., Azarbayejani, A., Darrell, T., Pentland, A.: Pfinder: Real-time tracking of the human body. IEEE Transactions on Pattern Analysis and Machine Intelligence, 780–785 (1997)
3. Fujiyoshi, H., Lipton, A.J., Kanade, T.: Real-time human motion analysis by image skeletonisation. IEICE Trans. Inf. & Syst. (January 2004)
4. Tao, Z., Ram, N.: Tracking multiple humans in complex situations. IEEE Transactions on Pattern Analysis and Machine Intelligence 26(9) (September 2004)
5. Agarwal, A., Triggs, B.: Recovering 3d human pose from monocular images. IEEE Transactions on Patten Analysis and Machine Intelligence 28(1) (January 2006)
6. Ayers, D., Shah, M.: Monitoring human behavior from video taken in an office environment. Image and Vision Computing (2001)
7. Yacoob, Y., Black, M.J.: Parameterized modeling and recognition of activities. In: ICCV (1998)
8. Jebara, T., Pentland, A.: Action reaction learning: Analysis and synthesis of human behavior. In: IEEE Workshop on the Interpretation of Visual Motion (1998)
9. Aggarwal, J.K., Cai, Q.: Human motion analysis: A review. In: Computer Vision and Image Understanding (1999)
10. Moore, D.J., Essa, I.A., Hayes, M.H.: Exploiting human actions and object context for recognition tasks. In: ICCV (1999)
11. Chomat, O., Crowley, J.: Probabilistic recognition of activity using local appearance. In: International Conference on Computer Vision and Pattern Recognition (CVPR), Vancouver, Canada (June 1999)
12. Intille, S., Bobick, A.: Recognizing planned, multi-person action. Computer Vision and Image Understanding (2001)

13. Hoey, J., Bertoldi, A.V., Poupart, P., Mihailidis, A.: Assisting persons with dementia during handwashing using a partially observable markov decision process. In: International Conference on Computer Vision Systems (ICVS), Germany (March 2007)
14. Ivanov, Y., Bobick, A.: Recognition of visual activities and interactions by stochastic parsing. IEEE Transactions on Patterns Analysis and Machine Intelligence (2000)
15. Vu, V., Bremond, F., Thonnat, M.: Automatic video interpretation: A novel algorithm based for temporal scenario recognition. In: The Eighteenth International Joint Conference on Artificial Intelligence, September 9-15 (2003)
16. Avanzi, A., Bremond, F., Tornieri, C., Thonnat, M.: Design and assessement of an intelligent activity monitoring platform. EURASIP Journal on Applied Signal Processing, Special Issue on Advances in Intelligent Vision Systems: Methods and Applications (August 2005)
17. Allen, J.F.: Maintaining knowledge about temporal intervals. Communications of the ACM (1983)

Remembering Pictures of Real-World Images Using Eye Fixation Sequences in Imagery and in Recognition

Geoffrey Underwood[1], Neil Mennie[1], Katherine Humphrey[1], and Jean Underwood[2]

[1] School of Psychology, University of Nottingham, Nottingham NG7 2RD, UK
{geoff.underwood,neil.mennie}@nottingham.ac.uk
lpxkah@psychology.nottingham.ac.uk
[2] Division of Psychology, Nottingham Trent University, Nottingham NG1 4BU, UK
jean.underwood@ntu.ac.uk

Abstract. Two experiments examined the eye movements made when remembering pictures of real-world objects and scenes, and when those images are imagined rather than inspected. In Experiment 1 arrays of simple objects were first shown, and eye movements used to indicate the location of an object declared as having been present in the array. Experiment 2 investigated the similarity of eye fixation scanpaths between the initial encoding of a picture of a real-world scene and a second viewing of a picture and when trying to imagine that picture using memory. Closer similarities were observed between phases that involved more similar tasks, and the scanpaths were just as similar when the task was presented immediately or after two days. The possibility raised by these results is that images can be retrieved from memory by re-instating the sequence of fixations made during their initial encoding.

1 Introduction

The eye movements made when looking at images reflects the purpose of inspection, with fixations made on regions of interest [1, 2, 3]. The question asked here is whether we can guide our eye movements volitionally as part of an image retrieval process. When attempting to remember a picture the sequence of eye fixations may match the sequence made during the initial inspection, and if this is the case, then it may be possible to retrieve a picture using eye movements as an interface for image retrieval. Oyekoya and Stentiford [4] have considered the possibility of using an operator's eye movements as a natural tool for content based image retrieval (CBIR), whereby image librarians would call up a required picture by imagining it and moving their eyes around a blank screen as they would when inspecting it. We consider the plausibility of using eye movements as a retrieval tool here with two experiments that examine the underlying processes that must be observed if such a CBIR interface can be implemented. The first experiment investigates the knowledge that an operator has about the locations of components of a display by asking for an indication of location by fixation of where a specific object was placed in the original image. In the second experiment we compare sequences of fixations – scanpaths – made during initial

B. Caputo and M. Vincze (Eds.): ICVW 2008, LNCS 5329, pp. 51–64, 2008.

inspection, during a recognition test, and during retrieval through imagery. The experiments are directed at the question of whether operators' eye movements when they imagine a picture can be used as indicators of successful retrieval.

Are sequences of fixations repeated when re-inspecting or imagining a picture? The order and pattern of fixations made by the viewer when looking at a scene has been described as a scanpath by Noton and Stark [5, 6]. Their theory predicts that the fixations made when first looking at a picture are very similar to those made when recognising the same picture at a later time. A number of studies have found that when participants view a picture for the second time, the scan patterns they produce are very similar to scan patterns produced on first exposure to the picture. For example, Foulsham and Underwood [7] presented a set of pictures in a recognition memory study, and then presented a second set to which an old/new decision was required. Scanpaths were similar when compared between two viewings of the same picture (encoding *vs.* old), and this similarity was greater than control comparisons (encoding *v* new and old *v* new). It is possible that we do not reproduce the same scanpaths over time due to the sequence of eye movements being stored internally or being related to an internal visual image, but that we do so because of the bottom-up influences of the visual stimulus. Our fixations are attracted by conspicuous or salient regions of the image [8, 9, 10], and when we are shown that same picture again at a later time, perhaps we simply look at the same regions of the picture again because those regions still possess the same low-level properties as when it was first inspected. Repeated scan patterns may be generated by viewers remembering how they inspected a picture when they first looked at it, but it could be that the features of the image - either bottom-up visual features or top-down meaningful features – are what drive the sequence of fixations.

Rather than comparing the fixation scanpath recorded during the initial inspection against the scanpath recorded during a recognition test and, as a consequence, risk finding a similarity due to an influence of image characteristics, we could use an imagery task. If scan patterns are then reproduced it cannot be due to external bottom-up influences, as no visual stimulus is present. Brandt and Stark [11] found substantial similarities between sequences of fixations made whilst viewing a simple chequer-board diagram and those made when imagining it later. Holsanova, Hedberg and Nilsson [12] used natural, real life scenes and found similar results. Visual input is not necessary to elicit repeated fixation patterns: when the imagery task is performed in the dark (i.e., without any possible visual features) eye movements still reflect objects from both the description and the picture [13]. Since there is no actual diagram or picture to be seen during the imagery period in these studies, it is likely that an internalised cognitive perceptual model is in control of these repeated scanpaths. In a modified version of the imagery experiment, Laeng and Teodorescu [14] found that viewers who fixated their gaze centrally during the initial scene perception did the same, spontaneously, during imagery. They also showed that viewers who were free to explore a pattern during perception, when required to maintain central fixation during imagery, exhibited decreased ability to recall the pattern. This is possibly because the oculomotor links established during perception could not be used in the process of building up a mental image, and this limitation impaired recall.

Eye movements are fundamental to Kosslyn's [15] visual buffer model of imagery, which is used to represent an internal image that conserves distance, location and orientation of the internal image. Eye movements during imagery would be associated with the internal shifts of attention around the representation in the visual buffer. The sequence of fixations made at encoding could be stored in the visual buffer as a spatial model, creating an internal representation of the visual stimulus. This scanpath is then reproduced at imagery to retrieve the internal representation.

2   Experiment 1: Retrieving an Object

When trying to imagine where an object was located, can we use eye movements? The task in this experiment was simple: viewers looked at an array of common everyday objects, and then identified the location of a target object by imagining where it was and then moved their gaze to that location. Successful use of eye movements to indicate the remembered location is necessary if voluntary gaze control is to be used as part of the process of retrieval through imagery.

2.1   Method

Nineteen university students participated in the experiment. All observers had normal or corrected to normal vision and were naïve as to the purpose of the experiment.

Participants were seated in front of a monitor that was viewed at a distance of 60cm. Their head was stabilised on a chin-rest. Displays were presented on a colour monitor, and eye movements were recorded with a head-mounted Eyelink II eye tracker sampling at 250 Hz and with an average degree of accuracy of <0.5 deg. Calibrations were performed prior to the start of each session, and an online drift correction was performed prior to the start of each trial.

Images consisted of a 5 x 5 grid of black squares on a white background with simple objects located within 7 of the resulting squares. Each of the 25 squares was 5 cm x 5 cm in size, with each side subtending approximately 4.6 deg. The central square was centred on the fixation point (approx 1 cm in diameter) in the centre of the screen (see Fig. 1). The 7 objects in the grid were images of everyday objects were chosen from the Tarr Lab Object Databank at http://www.tarrlab.org

The grid contained 25 squares, and each of the 24 non-central squares was a target square equally often. Each participant completed 48 trials in total (24 target present; 24 target absent), with each grid containing 7 objects. The second image (test image) contained one object located in the central square. The first task was to indicate whether the central test image had been presented in the grid.

Each trial started with a red fixation dot in the centre of an empty grid (see Fig. 1). The task was to maintain fixation on the red dot. After a 2 second interval, the red dot disappeared and the *first* image appeared, upon which they were then free to move gaze around the screen. The first image contained 7 objects, each placed in a separate square. They were asked to try and remember what the seven objects were and in which squares they were located. They were informed that they could do this in their own time (with a maximum of 20 seconds before being timed out), and that when they were able to remember what and where the objects were, they were to then press

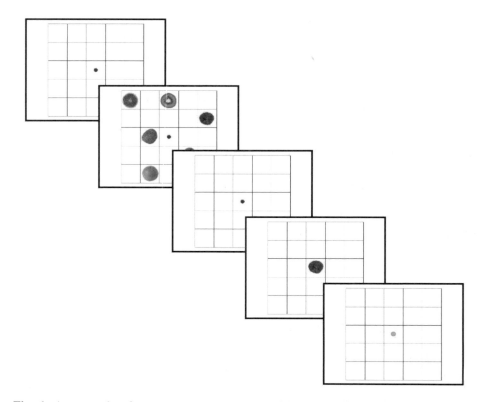

Fig. 1. An example of a target-present sequence of images used in Experiment 1. After inspecting the array of objects a single target object appeared in the centre of the 5 x 5 grid. The target object (a raspberry) appears on the right in a square in the second row of the grid. In this example the viewer has indicated correctly that the target was present, and a green fixation marker has appeared, prompting a re-direction of gaze to the location of the target in the original array. Images were presented in colour in the experiment.

a response button. The duration of the first image therefore varied from trial to trial as individuals tried to memorise the 7 objects and their locations.

After a 2 second interval, this dot was replaced by a *second* image containing only one image of an everyday object, and this object was located in the central square. Assuming that the participants were maintaining fixation on the dot then the single image would therefore appear at the locus of gaze. The participants then had maintain fixation on the object and to press one of two buttons (YES or NO) on the EyeLink button box to signify if the object had been present in the original set of 7 objects. If they pressed NO, then the trial ended. If they pressed YES, the image was replaced by a green fixation dot in the centre of the screen. The green dot was their signal to move gaze as accurately as possible over the blank grid to the square where they thought the object had been, and to then fixate within that square and to press the YES button again. The trial terminated once the YES button had been pressed for the second time, and so the duration of the blank screen with the green dot varied from trial to trial as participants "searched" for the correct square. Fig. 1 illustrates a typical trial sequence. All movements of gaze were recorded during a trial. Immediately prior to the session, participants undertook a practice session of 5 trials.

To establish how accurate the participants were when saccading from the centre to the correct target location on the blank screen, an "ideal" angle was calculated from the fixation dot to the centre of each of the 24 squares. This angle was then compared against the angle of the first saccade executed after the onset of the green dot that left the fixation square and landed in another square.

2.2 Results

The mean time taken by the 19 participants to inspect the first image during memorisation was 11.61 s. Participants were successful at recalling whether the object had been present or absent, with 80% of the target-present trials and 94% of the target-absent trials gaining correct responses. The mean time to correctly decide if the sole object at fixation had been was 1928 ms and to respond correctly that it was absent the mean was 1375 ms. A paired samples t-test indicated a reliable difference ($t = 9.02$, df = 37, $p < 0.001$). This is a curious result, and it might be expected that viewers would be faster when they know that a target is present. One possibility here is that knowledge of the second decision acts to slow them down. When they indicated that a target was present they knew that they would be asked a subsequent question about its location, and this processing of location information may have been initiated early, and interfered with the execution of the first decision. (In a subsequent experiment we eliminated this stage, and obtained improved responses – see below.) Participants get a higher percent correct when the object was absent from the original set, and they also respond faster.

If the participants pressed the YES button, then the blank screen with the green fixation dot would appear. At this point they had to move gaze across the grid to the square where they thought the target object had been located and to then press the YES button a second time while fixating within that square. We looked at the time taken to perform this task, the accuracy of recall (i.e. how often they looked at the correct square), and we also looked at how fast and accurate their first saccade was when initiating this search.

The mean time taken to locate a square (and to press the button) when the target had been present was 2830 ms. This was similar to the few trials (n=30) where the response was a false positive (mean of 2884 ms).

The spatial memory for target locations, as evidenced by the ability to direct gaze to the correct square, was much poorer than the memory for the object itself. Fixation was only directed at the correct square on 47% of occasions (i.e. that square was actually the target 47% of the time), which is much lower than the 80% accuracy for recall shown above. We also looked at each *initial* saccade that left the central square and landed in a separate square of the blank grid, and calculated the angle (out of 360 degrees) of the saccade and its difference to the "ideal" angle (a straight line between the fixation dot and the centre of the target square), its amplitude and its peak velocity.

The mean difference (or error) between first saccades on all trials that correctly fixated the target, relative to the ideal trajectory, was 12 deg, while for those trials where the incorrect square was identified the mean error was 61 degrees (t = 11.21, df = 18, p < 0.001). These differences are absolute values, meaning that for those saccades that accompanied correct decisions about location, the saccades were likely to fall within a 24 deg corridor around the "correct" angle for that target square. This is illustrated in Fig. 2 below, with the 122 deg angle illustrating the mean difference between initial saccades to incorrectly identified squares and the correct angle for that trial's target.

There was also a significant difference in the peak velocity of initial saccades that were directed to the target, with saccades of correct trials having a peak velocity of 363 deg/s^2 and saccades that initiated gaze towards incorrect locations having a peak velocity of 333 deg/s^2 (t = 2.56, df = 18, p = 0.02). Viewers make more direct saccades to the target when they get it right (i.e. they are not often going off in another direction before returning to the correct square), and they also make significantly faster eye movements on the correct trials.

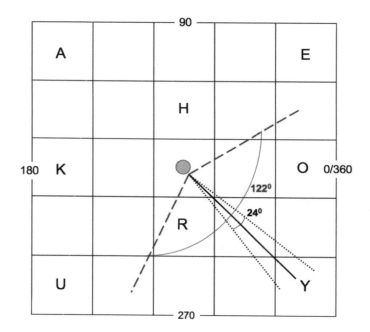

Fig. 2. The mean difference between the direction of the initial saccade and a straight line towards a target in square Y as an example. The angles of all *initial* saccades that left the central square were calculated, and the difference between these and the straight line to the target determined. For those trials where the correct target square was fixated, this "error" was 12 deg. As the error could be on either side of the straight line, we illustrate this as a 24 deg angle around the line (fine dashed lines). For those trials where the incorrect square was fixated at button press the error was 61 deg, here illustrated as a 122 deg angle (thick dashed line). When there was no decision about whether the target had been presented, because there was a target present on every trial, the saccadic "error" was reduced to 9 deg.

In a subsequent version of the experiment, with a new sample of 19 participants drawn from the same population as used here, we eliminated the decision as to whether the target object had been presented in the array – all trials presented targets that had been presented. Viewers had only to move their eyes to the position in the array where the target had appeared. The correct location was selected on 70% of occasions when this procedure was used, and the accuracy of the initial saccadic movement improved to 9 deg, relative to the 24 deg when there was an intervening decision about whether the target had been presented.

3 Experiment 2: Retrieving a Picture of a Real-World Scene

The aim of this experiment is to investigate whether sequences of eye fixations are similar during a recognition test and during an imagery test, relative to the sequence during initial inspection. Viewers were again required to recognise the images inspected, and then an imagery task was conducted. One of the aims of current experiment is to determine whether scanpaths at encoding and imagery are stable over extended periods of time. If fixation scanpaths are to be usable as CBIR aids they would need to be stable over time, and so after a two day interval the imagery task was repeated.

3.1 Method

Fifteen university students took part in this experiment. All participants had normal or corrected-to-normal vision. Inclusion in the study was contingent on reliable eye tracking calibration and the participants being naïve about eye movements being recorded.

Eye position was recorded using an SMI iView X Hi-Speed eye tracker, which uses a tower-based ergonomic chinrest and provides gaze position accuracy of 0.2 deg, with a sampling rate of 240 Hz. This eye tracker records eye position non-intrusively. A set of 60 high-resolution digital photographs of real-world scenes were used as stimuli, sourced from a commercially available CD-ROM collection ("Art Explosion"). Each picture was distinctly individual, in that given a short description it could be not mistaken for any of the others: in a pilot study a further 10 participants correctly matched 100% of the pictures and their intended labels. The labels were used during the imagery phases of the experiment. Half the pictures were designated as "old" and shown in both encoding and test phases, while the other half were labelled "new" and were shown only as fillers at test. Pictures were presented on a colour computer monitor at a fixed viewing distance of 98 cm that gave an image that subtended 25.03 by 18.83 deg. Examples of two of the images used, and their labels, are shown in Fig. 3.

Participants were told that their pupil size was being measured in relation to mental workload. The first phase of the study involved viewing a set of 30 pictures, presented in a random order, in preparation for a memory test. Each picture was preceded by a fixation cross for 1 s, which ensured that fixation at picture onset was in the centre of the screen. Each picture was presented for 3 s, during which time participants moved their eyes freely around the screen.

Fig. 3. Examples of images used in Experiment 2. All images could be identified by a single word or short phrase – "the American football game" and "the penguins" in the case of the two images here. Images were presented in colour in the experiment.

After all 30 pictures had been presented, participants saw a second set of pictures and had to decide whether each picture was new (not seen before in the experiment) or old (from the previous set of pictures), pressing N or O on the keyboard if the picture was new or old. Sixty stimuli were presented in a random order, 30 of which were old and 30 new. Each picture was again shown for 3 s and participants could only make a response after this time. This was to encourage scanning of the whole picture.

After all 60 pictures in the recognition test had been shown, the participants took a break before performing an imagery task. This time they saw 30 white screens with a short sentence describing one of the pictures they had just seen. All the pictures in this imagery task had been seen previously. The pictures appeared in a random order. Participants were asked to imagine the picture described and try to remember everything they could about it. Each description appeared for 3 s and then the screen went blank for 5 s, in which time they attempted to imagine the stimulus.

Participants returned two days later to perform the last imagery task again. The procedure was identical and all of the descriptions of pictures in this task had previously appeared in the first imagery task, and were presented here in a new random order.

3.2 Results

In the following analyses data were excluded from trials where the fixation at picture onset was not within the central region or when calibration failed. Recognition accuracy was very high (97.11%) and no further analysis was performed on this measure. The measure of eye fixation behaviour for this experiment used string analyses for the comparisons of sequences of fixations.

Fixation sequences were analysed using string editing, to identify the similarity between sequences produced on encoding and imagery, encoding and recognition, encoding and delayed imagery, imagery and recognition, imagery and delayed recognition, and recognition and delayed imagery. This string editing technique has been described in detail elsewhere [7, 11, 16, 17], and involves turning a sequence of fixations into a string of characters by segregating the stimulus into labelled regions.

The similarity between two strings is then computed by calculating the minimum number of editing steps required to turn one into the other. An algorithm for calculating the minimum editing cost was then applied to derive a single value that represents the similarity between two strings or scanpaths [7, 11].

For each participant, the scanpath recorded on first viewing each image was calculated, and compared against that participant's viewing during the recognition task. This comparison delivered a string similarity value for each picture, and these values were then averaged to generate a single value for each participant. This process was repeated so that we could also compare initial encoding against initial imagery, against delayed imagery, and to compare initial imagery against delayed imagery. The results of the comparisons are shown in Fig. 4, with perfect similarity having a value of 1. Sequences of eye movements were less similar when comparing encoding and imagery then when comparing encoding and recognition. We compared string similarities between the two encoding and imagery phases against string similarities between encoding and recognition. An ANOVA showed a reliable effect of string comparison type ($F(3,42) = 12.66$, $p<0.001$).

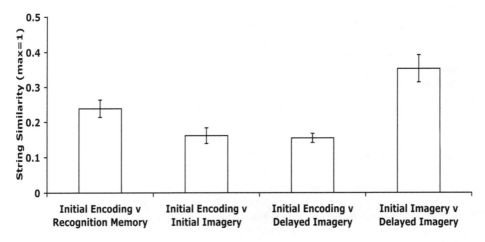

Fig. 4. Differences in fixation sequence similarities (means and SEMs), using Levenshtein's string-editing method, between the encoding, imagery, and recognition phases.

Greatest similarity is seen between the fixation sequences made during the two imagery phases, and moderately high similarity between the sequences during encoding and recognition, but all four comparisons are reliably greater than the similarity score expected by random fixations. To determine the string similarity that would be expected by chance we matched each scanpath from an individual participant against that obtained with a different picture viewed by the same participant. The comparison picture for each scanpath was selected randomly. Averaging over all pictures and over all participants, this procedure generates a string similarity by chance of 0.00742, where the maximum is again 1.

The similarity scores shown in Fig. 4 were compared against each other. Post-hoc comparisons showed that there were reliable differences comparing the string similarity scores between *initial encoding v initial imagery* and *initial encoding v recognition memory* (p<0.05), with more similarity between scan patterns involving recognition than imagery. There was also a difference between *initial encoding v initial imagery* and *initial encoding v delayed imagery* (p<0.05). The similarity between fixation sequences during encoding and recognition, when the image was available for inspection, was greater than that between encoding and either of the imagery tests. The similarity between *initial imagery v delayed imagery* was also greater than either of the similarity scores involving encoding and imagery: *initial imagery* (p<0.001) and *delayed imagery* (p<0.001). There was also greater similarity between the two imagery tests than between encoding and recognition (p<0.01).

4 Discussion

Participants in both experiments were very good at identifying images as old or new. The accuracy rate was so high because each picture had to be distinctly individual in order for the imagery and delayed imagery tasks to work. This made it easy to decide which pictures had been seen before and which had not.

Average fixation duration were measured and analyses found a main effect of group in that participants in Task 1 made shorter fixations than participants in Task 2. Average fixation duration at encoding was almost identical for Task 1 and Task 2; suggesting that the groups were well matched and the differences between groups in other conditions were affects of the experimental design. Participants in Task 1 were asked to imagine the picture soon after they has seen the visual stimulus, whereas in Task 2 participants saw all the stimuli at once, then were given a recognition memory test before an imagery task. So, the pictures in Task 1 should have been easier to imagine as they were still in short term memory and there was no decision to be made about which picture should be retrieved – it was always the picture most recently seen. Therefore participants may have been able to remember where different parts of the scene were without having to move their eyes as much. Additionally, participants in Task 1 were given a recognition memory test in between encoding and imagery, which involved twice as many stimuli than were in the imagery task, so the confusion due to an increased memory load could have resulted in longer fixation times while participants tried to retrieve the correct image. This could be argued as support for Kosslyn (1994) in that the increase in working memory load makes it harder to access internal information from the 'visual buffer'. Furthermore, the immediate imagery phase in Task 2 was more difficult as it involved only seeing a word that described a picture rather than an image itself, it may have taken longer to recall what features of the picture were at a certain point of fixation. This may also have resulted in longer fixation durations during the imagery attempts.

The average fixation durations in both Tasks during the first imagery attempt were almost identical to those when the imagery attempt was repeated two days later. Laeng and Teodorescu (2002) found that eye movements when participants first saw the picture were very similar to those at imagery and that if participants who moved their eyes freely in encoding were then made to keep a central fixation at imagery, recall performance decreased dramatically. This suggests that eye movements at first viewing help to encode the picture and reproducing those eye movements at a later stage may help recall that picture. More recent research has also found that scan patterns are also stable over multiple viewings (Humphrey & Underwood, *submitted*). So, one hypothesis for the similarity of average fixation durations on both Tasks at imagery and delayed imagery is that eye movements at imagery become associated with the memory of the picture so that in the delayed imagery condition, eye movements are very similar to the first imagery task because they act as cues that can be used in retrieval. This supports the idea of a 'visual buffer' to the extent that it seems possible to retrieve an internal image from working memory; however, these results suggest that if such a mechanism does exist, it must be flexible. Otherwise it would be expected that eye movements at encoding would be the same as those at recognition and imagery and delayed imagery, which they are not. At recognition, participants in Task 1 may have made shorter fixations because they had 'inspected' each picture twice before the recognition test (once during encoding and once during imagery) so recognition may have been easier and less exploration of the picture was needed.

In both Tasks, participants made shorter fixations at encoding and at recognition than at imagery or delayed imagery. At encoding, the participants saw the pictures for the first time and therefore tried to take in as much information as possible in the 3 seconds available. The average fixation durations of 350 ms for Task 1 and 374 ms for Task 2 at encoding are consistent with previous studies. At recognition, participants saw the pictures for a second time. Our previous study also found a high similarity between scan patterns at encoding and recognition [7]. The finding that average fixation duration is similar at encoding and recognition (averaging over both Tasks) suggests that the visual buffer may store temporal information as well as the spatial pattern of eye movements.

Scan patterns produced at each condition were compared to every other condition using string analysis to create a similarity score. In Task 1 (imagery first), scan patterns were more similar when comparing imagery and delayed imagery than when comparing encoding and imagery or encoding and delayed imagery or imagery and recognition. This could be explained in terms of mixed and pure process comparisons. When comparing imagery and delayed imagery, the task was the same in Task 2 and very similar in Task 1, in that both conditions involved recalling a memory without any immediate visual cues. This could be referred to as a 'pure process comparison'. Whereas when comparing encoding and imagery or encoding and delayed imagery or imagery and recognition, one of the conditions in each comparison involved visual input from the stimulus and the other involved recalling without any visual input. These could be referred to as 'mixed process comparisons', and produce lower similarity scores.

The most similar scan patterns came from pure process comparisons where there was similar visual input in each condition (imagery compared to delayed imagery and encoding compared to recognition), and from comparisons that mimicked the same retrieval processes (imagery compared to recognition, and delayed imagery compared to recognition). The lowest scan pattern similarity scores came from mixed process comparisons (encoding compared to imagery, encoding compared to delayed imagery, and imagery compared to recognition in Task 1). The more similar the retrieval process is to the encoding process, the more similar the scan patterns produced. This provides evidence in favour of the use of a visual buffer, as suggested by Kosslyn [15], in that visual information is stored in working memory, but it also suggests that the model may be more complicated than simply shifting attention to different parts of an internal image. The relationship between the encoding and retrieval process seems to be very important and one might even suggest the existence of facilitatory and inhibitory pathways within the model.

Propositional accounts such as that of Pylyshyn [18] argue that there is no such thing as the visual buffer and that when participants are asked to "imagine X" they use their knowledge of what "seeing X" would be like, and they simulate as many of these effects as they can. However, it seems very unlikely that participants are able to mimic a behaviour so precisely in their eye movements. In agreement with Johansson et al [13] the number of points and the precision of the eye movements to each point are too high to be remembered without a support to tie them together in a context, such as an internal image. This is further supported by the finding that temporal information as well as spatial information is reproduced at retrieval and is consistent over time as long as the same retrieval process is used. Furthermore, if participants did store spatial scene information as a large collection of propositional statements, scan pattern similarity should have remained constant across conditions despite changing the retrieval task, but this was not the case.

The finding that scan patterns at imagery were highly similar to those at delayed imagery (48 hours later) suggests that they are stable over time. This challenges the suggestion that the sensorimotor trace is stored only in short-term memory. The similarity between the scan patterns also lends support for Scanpath Theory [18], which suggests that since there is no actual diagram or picture to be seen during the imagery period, it is likely that an internalised cognitive perceptual model is in control of these scan patterns.

Could eye movement scanpaths be used in the retrieval of images (CBIR) from large pictorial libraries? For such an interface to operate, the viewer's eye movements when attempting to retrieve an picture must be distinctive in that they should identify the picture being imagined. The scanpath would not need to be unique, but the interface would be most functional if a scanpath delivered only a small number of candidates from the library. The present experiments give some reason to believe that a scanpath-based interface might be workable. In the first experiment we established that viewers have good location memory when looking at arrays of objects, and in the second experiment we compared the scanpaths made when first viewing a picture against both a later viewing and later imagery tests. The similarity between scanpaths did not approach unity, but for each comparison it was reliably better than chance would predict. Scanpaths may be sufficiently distinctive to identify the associated image. Cerf, Harel, Huth, Einhäuser and Koch [20] have recently presented evidence that scanpaths are not only distinctive but that they can indeed be used in a discrimination task. When viewers are presented with a set of images and with a set of scanpaths from one viewer looking at those images, then matching the scanpath with its image can be performed reliably better than chance. People can identify the scanpath that is derived from a specific picture.

To conclude, in accordance with Johansson et al [13] the results of this paper lend support to Kosslyn's [15] visual buffer model of imagery, and challenge Pylyshyn's [18] propositional visual index model. The variations in scan pattern similarities caused by manipulation of the retrieval processes suggests that the visual buffer may be more complicated than previously thought, with possible facilitatory and inhibitory pathways, and also that scan pattern theory and the visual buffer may be linked. The sequence of fixations made at encoding are stored in the visual buffer as a spatial model, creating an internal representation of the visual stimulus. This scan pattern is then reproduced at imagery to retrieve the internal representation. The use of eye fixation scanpaths as an interface for the content based image retrieval (CBIR) of pictures from libraries remains a viable possibility, given that viewers can direct their eyes to indicate the previous locations of objects in scenes, and that scanpaths when imagining a scene are reliably similar to those made during initial inspection. Cerf et al. [20] have also demonstrated that viewers can identify a picture from a small set of candidates on the basis of a scanpath made in response to that picture, again demonstrating the viability of using scanpaths in a CBIR interface. The absence of a perfect match between scanpaths at encoding and during imagery indicates that when retrieving the required picture a scanpath-based interface may deliver a set of candidates rather than the unique picture that is required.

Acknowledgment. We are grateful to the EPSRC (UK) for supporting this work with project award EP/E006329/1.

References

1. Yarbus, A.L.: Eye Movements and Vision. Plenum, New York (1967)
2. Mackworth, N.H., Morandi, A.J.: The gaze selects informative details within pictures. Percept. and Psychophysics 2, 547–552 (1967)
3. Underwood, G., Jebbett, L., Roberts, K.: Inspecting pictures for information to verify a sentence: Eye movements in general encoding and in focused search. Quart. J. Exp. Psychol. 57A, 165–182 (2004)
4. Oyekoya, O.K., Stentiford, F.W.M.: Eye tracking as a new interface for image retrieval. BT Tech. J. 22, 161–169 (2004)
5. Noton, D., Stark, L.: Scanpaths in saccadic eye movements while viewing and recognizing patterns. Vis. Res. 11, 929–942 (1971a)
6. Noton, D., Stark, L.: Eye movements and visual perception. Sci. Am. 224, 34–43 (1971b)
7. Foulsham, T., Underwood, G.: What can saliency models predict about eye movements? Spatial and sequential aspects of fixations during encoding and recognition. J. Vis. 8(2):6, 1–17 (2008)
8. Itti, L., Koch, C.: A saliency-based search mechanism for overt and covert shifts of visual attention. Vis. Res. 40, 1489–1506 (2000)
9. Underwood, G., Foulsham, T., van Loon, E., Underwood, J.: Visual attention, visual saliency, and eye movements during the inspection of natural scenes. In: Mira, J., Alvarez, J.R. (eds.) IWINAC 2005. LNCS, vol. 3562, pp. 459–468. Springer, Heidelberg (2005)
10. Underwood, G., Foulsham, T.: Visual saliency and semantic incongruency influence eye movements when inspecting pictures. Quart. J. Exp. Psychol. 59, 1931–1949 (2006)
11. Brandt, S.A., Stark, L.W.: Spontaneous eye movements during visual imagery reflect the content of the visual scene. J. Cog. Neuro. 9, 27–38 (1997)
12. Holsanova, J., Hedberg, B., Nilsson, N.: Visual and verbal focus patterns when describing pictures. In: Becker, W., Deubel, H., Mergner, T. (eds.) Current Oculomotor Research: Physiological and Psychological Aspects, pp. 303–304. Plenum, New York (1998)

13. Johansson, R., Holsanova, J., Holmqvist, K.: Pictures and spoken descriptions elicit similar eye movements during mental imagery, both in light and in complete darkness. Cog. Sci. 30, 1053–1079 (2006)
14. Laeng, B., Teodorescu, D.-S.: Eye scanpaths during visual imagery reenact those of perception of the same visual scene. Cog. Sci. 26, 207–231 (2002)
15. Kosslyn, S.M.: Image and Brain. MIT Press, Cambridge (1994)
16. Privitera, C.M., Stark, L.W.: Algorithms for defining visual regions-of-interest: Comparison with eye fixations. IEEE Trans. Patt. Anal. Mach. Intell. 22, 970–982 (2000)
17. Levenshtein, V.: Binary codes capable of correcting deletions, insertions and reversals. Soviet Physice - Doklady 10, 707–710 (1966)
18. Stark, L., Ellis, S.R.: Scanpaths revisited: cognitive models direct active looking. In: Fisher, D.F., Monty, R.A., Senders, J.W. (eds.) Eye Movements: Cognition and Visual Perception, pp. 193–227. Erlbaum, Hillsdale (1981)
19. Pylyshyn, Z.W.: Mental imagery: In search of a theory. Behav. Brain Sci. 25, 157–238 (2002)
20. Cerf, M., Harel, J., Huth, A., Einhäuser, W., Koch, C.: Decoding what people see from where they look: Predicting visual stimuli from scanpaths. In: Paletta, L., Tsotsos, J.K. (eds.) Proceedings of the 5th International Workshop on Attention in Cognitive Systems, Santorini, Greece, pp. 194–206 (2008)

Towards a Model of Information Seeking by Integrating Visual, Semantic and Memory Maps

Myriam Chanceaux[1], Anne Guérin-Dugué[1], Benoît Lemaire[1],
and Thierry Baccino[2]

[1] University of Grenoble, France
`<first name>.<last name>@imag.fr`
[2] University of Nice-Sophia-Antipolis, France
`baccino@unice.fr`

Abstract. This paper presents a threefold model of information seeking. A visual, a semantic and a memory map are dynamically computed in order to predict the location of the next fixation. This model is applied to a task in which the goal is to find among 40 words the one which best corresponds to a definition. Words have visual features and they are semantically organized. The model predicts scanpaths which are compared to human scanpaths on 3 high-level variables (number of fixations, average angle between saccades, rate of progression saccades). The best fit to human data is obtained when the memory map is given a strong weight and the semantic component a low weight.

Keywords: computational model, information seeking, visual saliency, semantic knowledge, memory.

1 Introduction

Over the past decade, a large amount of writings or contents has become available to the web user. However in the same time, there has been relatively limited progress towards a scientific understanding of the psychology of human interaction with the web. Detailed integrated cognitive models are difficult to create, limited to very narrow experimental conditions of interaction and mostly unable to face with the semantics of web contents.

One of the most frequent tasks on the web consists in seeking information on pages. Searching for information requires defining a given goal that may be precise or vague, more or less variable along the navigation. When the goal is well-defined, top-down models like ACT-R [1] predict relatively well how the information is retrieved on the web [2]. When the goal is ill-defined, the user must rely on data displayed on the page and incrementally build/maintain into memory the goal to reach information. Such task requires substantial acquisition and integration of knowledge coming from external sources [3] in order to better define goals, available courses of action, heuristics and so on. Modeling in this case must take into account perceptual information carrying out mostly bottom-up processes that analyze data presentation along the visual exploration of pages

B. Caputo and M. Vincze (Eds.): ICVW 2008, LNCS 5329, pp. 65–78, 2008.

and guide the user attention. However, this visual processing is closely related to content processing and any computational model (psychologically valid) has to explain how this integration of information is generated on-line to verify whether the goal is reached or not.

While a number of studies have investigated this activity, very few cognitive models are satisfying. Some models are very general pointing out the activity of information seeking [4,5,6] but saying nothing or very few about the underlying cognitive processes. Others are too specific dedicated to web navigability as COLIDES [7] or some extensions of ACT-R model such as SNIF-ACT [8] or BSM [9]. The scope of the paper is to sketch an integrated cognitive model that account for both visual and semantic processing during information seeking.

2   Information Seeking: 3 Components

Seeking information (i.e, a word) in a document requires from the user to process two sources of information: visual information (i.e, exogenous information) and semantic information (i.e, endogenous information). The former refers to low-level visual features involving bottom-up selective processes while the latter refers to word meaning represented in semantic memory and entails top-down processes. Both visual and semantic information have been shown to guide the visual scanpath, the gaze tends to move towards locations that are visually salient, but it is also attracted to regions that are semantically relevant with respect to the current search goal. Let's describe more precisely this respective influence.

2.1   Visual Information

Computational models of selective visual attention have attracted growing levels of interest during this last decade. The purpose is to predict where humans look when they perform a visual detection task from a bottom-up perspective. Most of these models are mainly based on two original concepts: the Feature Integration Theory [10]. Among them, the most popular is proposed by Itti and Koch [11]. It is based on a feature decomposition of the visual stimuli (natural visual scenes), and a competition-fusion process between the parallel feature maps that extracts a visual saliency map. The highest salient regions are then segmented and sorted according to their saliency value. For the first eye fixations on a picture, these models fits well the eye movements data when the visual stimuli have little semantic information and when the task is free without explicit task driving the scene exploration [12]. In the case of more demanding visual search, the visual saliency is progressively modulated over time by semantic and cognitive controls, depending of the type of the scene (a priori knowledge of the scene) and the task [13]. See for example the discussions in [14]. Due to this complexity, few models integrate these two pathways. Among recent propositions, [15] extracts saliency regions through interactions with a working memory and a semantic and visual long term memory. In our context of information seeking, visual stimuli can be considered as more simple than natural scenes from the point of view of the image features, but very complex from the point of view of the meaning and the semantic of the scene. Thus these bottom-up models must be highly simplified in the final model, but they must integrate specificities of reading tasks as it is proposed in [16].

2.2 Semantic Information

Looking for a word entails also the automatic access to semantic memory. Only very few computational models of information seeking have taken into account this top-down process probably due to the difficulty to represent meaning for a computer. However, since the development of Latent Semantic Analysis [17], meaning representation can be computed and estimated. Basically, LSA takes a huge corpus as input and yields a high-dimensional vector representation for each word, usually about 300 dimensions. It is based on a singular value decomposition of a word × paragraph occurrence matrix, which implements the idea that words occurring in similar contexts are represented by close vectors. Such a vector representation is very convenient to give a representation to sentences that were not in the corpus: the meaning of a sentence is represented as a linear combination of its word vectors. Therefore, we can virtually take any sentence and give it a representation. Once this vector is computed, we can compute the semantic similarity between any word and this sentence, using the cosine function. The higher the cosine value, the more similar the words are.

One of the first models attempting to explain information seeking by using LSA was COLIDES (Comprehension-based Linked model of Deliberate Search) [7]. It describes how people attend to and comprehend information patches on individual webpages. It is a simulation model of navigation trying to extend a series of earlier models developed by Kitajima & Polson and the Kintsch's construction integration theory of text comprehension. In COLIDES, the description of a web page is made up of a large collection of objects competing for users' attention, which are meaningful units and/or targets for action. Users manage this complexity by a two-phased processes: 1) an Attention Phase where users segment the page into regions and focus on a region of the page; 2) an Action Selection Phase in which users first comprehend each of the objects (e.g., hypertext link, graphic, radio button, etc.) that can be acted on in the focused-on region, including the consequences of acting on the object. Then they select one of the actions, usually clicking on one of the available hyperlinks. In both phases, the user's behaviors are determined by the perceptions of semantic similarity between the user's goals and the descriptions of alternative regions or actions. This similarity is calculated by LSA. Despite this interesting semantic component, COLIDES does not describe precisely how low-level information coming from vision or attention processes can guide the user gaze and orient the selection.

2.3 Memory Mechanism

A model of selective visual attention and scanpaths would be incomplete without describing the process, by which the currently attended location is prevented from being attended again, this mechanism is known as the Inhibition of Return (IOR). The IOR refers to an increased difficulty of orienting to a location to which attention has previously been directed. This possibly ensures that fixations are less likely to return to a previous point of high salience [18]. Pratt and Abrams [19] reported that the IOR of attention is found only for the most recently attended of two cued locations but other has shown that in more complex environments more possible locations may be involved [20].

3   Model

Our model is a spatio-temporal model based on the dynamic integration of visual and semantic information associated to a simple memory mechanism. It aims at describing in a cognitively plausible manner how visual, semantic and memory processes interact in order to predict eye movements in an information-seeking task. Each of these three components of our model is implemented by means of a conditional heat map of the current image, in which each of its elements is assigned a weight representing its relevance for the given component. Each of these maps is conditional on the location of the current fixation. The memory map is also conditional on the entire scanpath. These three maps are therefore continuously updated during the simulated visual search. Basically, they work in the following way: - the visual component integrates both the specific behavior of the human retina and the visual properties of the scene. Therefore, this component gives high weights to the fovea, but also to visually salient elements; - the semantic component gives high weights to the current zone if its semantic similarity with the goal is high, since the solution might be close; however, if this similarity is low, the current zone elements are assigned low values, meaning that is it probably not an interesting area; - the memory component strongly decreases the weight of the previous fixation zone, in order not to move back to it. However, since human memory is limited, these values tend to return to normal values over time. The three maps are integrated by a weighted sum and the simulated gaze is moved towards the best-weighted zone. Once the new fixation has been selected, maps are updated accordingly, then a new fixation is chosen, and so on. From an initial fixation point, our model thus produces a scanpath.

3.1   Task

Our final goal is to apply this model to complex web pages in which visual and semantic information are highly salient and generally not congruent. However, we first implemented and tested this general model on a simple task in order to control parameters as much as possible. Therefore, we largely simplified a general web page to only keep minimal visual and semantic data. We ended up with images containing 40 independent words (Fig. 1).

User goal. We formalized the goal the user is pursuing by considering that this user is seeking a particular piece of information. This item is defined by *the class it belongs to* and its specific features within this class. For instance, the user may

Fig. 1. Example of image. Instruction is: *quel est l'aliment le plus sucré ? (what is the sweetest food?)*. Correct answer is *confiture (jam)*.

look for *a scientific publication* given its title, *a tennis result* for a specific player, a *restaurant* that is open on Sunday, etc.

Our experimental users are thus instructed to find a specific word in the image which belongs to a given *category* and is the best at satisfying a given *feature*. This question is linguistically expressed by such a sentence: *find the most [feature] [category]*; for example, *find the most alcoholized beverage* or *find the roughest sport*. In the remainder of this paper, this user goal will be called *instruction* to avoid any confusion with the word the user should find.

Visual and semantic features of words. Each of the 40 words of each image has a visual feature and a semantic feature.

For the moment, the only visual feature of words is their font size, from 13 to 19pt. In a future experiment, words will be also characterized by colors.

We also organized the 40 words in order to reproduce a very common property of our world which is that objects that are similar to each other tend to be near each other. This semantic-spatial congruency helps us a lot when we search information: in newspapers, football and tennis results are close to each other; in supermarket, all vegetables are in the same place and they are close to fruits.

In each of our images, 7 words belong to the same category, including the target word. For instance, there are 7 alcoholized beverages in the first image mentioned previously, 7 names of sport in the second one, etc. All 33 other words are of decreasing semantic similarity with the instruction.

Figure 1 presents such an image, the instruction being *Find the sweetest food*. The target is *confiture* (jam). The six words that belongs to the same category are *citron* (lemon), *crème* (cream), *salade* (salad), *viande* (meat), *soupe* (soup) and *chocolat* (chocolate). Close to the target are also words related to food, like *saveur* (flavour), *marmite* (cooking-pot), *litre* (liter), but the more distant words are from the target, the less similar they are.

We tried to be as much objective as possible in the design of these images. To that end, we created a semantic space by applying Latent Semantic Analysis (LSA) [17] to a 13 million-word French corpus composed of novels, newspaper articles and encyclopedia chapters. Many articles in the literature have shown high correlations between LSA cosines and human judgements of similarity [21]. We thus relied on this LSA semantic space to define our 18 images. The procedure was the following:

- 18 [feature]-[category] instructions were defined;
- for each instruction, semantic similarities between the instruction and all words whose length was between 5 and 9 letter long and LSA weight was between .3 and .7 (medium frequency) were computed;
- the 7 best words which belonged to the category were selected;
- 33 other words were randomly selected at regular intervals between a 0 semantic similarity and the semantic similarity of the 7th previous word. For instance, the list corresponding to the instruction *sport brutal (rough sport)* is the following (similarity values are in parentheses):

1. football (soccer) (.71)	8. gardien (goalkeeper) (.37)
2. rugby (rugby) (.69)	9. victoire (win (.36)
3. tennis (tennis) (.57)	10. vainqueur (winner) (.35)
4. basket (basketball) (.48)	...
5. cyclisme (cycling) (.43)	37. charbon (coal) (.02)
6. course (run) (.42)	38. chêne (oak) (.01)
7. voile (sailing) (.38)	39. domicile (home) (.01)

- semantic similarities between all pairs of words were computed and a Multi-Dimensional Scaling procedure was run to assign all words 2D coordinates;
- all coordinates were scaled in order to fill an entire 1024x768 screen;
- in order to avoid word overlapping, 80 non-overlapping positions (NOP) were randomly defined and each word was moved to a close NOP such that the sum of these moves was minimum.

In order to select targets and be more precise about the 7 first words which play an important role in the task, we controlled their semantic similarity with the instruction by asking 28 participants to assess their similarities with the instruction on a 5-point scale. The target was defined as the word which was best-rated. We also computed a Student test between the first and second best-rated words to make sure there was no ambiguity on the target. In 4 cases out of 18, this difference was not significant. We then removed the second best-rated and replaced it by a word which was not highly similar to the instruction.

Experimental conditions. Last but not least, in order to investigate the relative contribution of semantic versus visual factors, we defined three visual conditions and two semantic conditions. Semantic conditions are (1) semantic organization of words as defined previously; (2) no semantic organization : words are randomly assigned to the locations of the previous condition.

Visual conditions are (1) random assignment of visual features to words from font size 13 to 19; (2) no visual features at all; (3) visual features are congruent to spatial locations: words that are close to the target have higher font size.

We therefore ended up with 108 images: 18 instructions x 2 semantic conditions x 3 visual conditions. We now present how we implemented our visual, semantic and memory maps to predict scanpaths on these images.

3.2 Maps

The unit of our maps should normally be the pixel, but for the sake of psychological validity, we are currently using the word since our images contain nothing but words. Actually, users are not looking for a region of pixels but for an object, here a word. Let us take an example. Suppose our model has already made 7 fixations on an image from which the instruction was: *find the most dangerous fish*. It is now fixating the word *baleine (whale)*.

Figure 2 displays the 3 maps corresponding to this scanpath as well as the integrated map. Word colors represent weights: the darker, the higher. When summing up the 3 maps, words are given new weights. The next fixation in made on the word that obtained the highest weight (*requin (shark)* in our example).

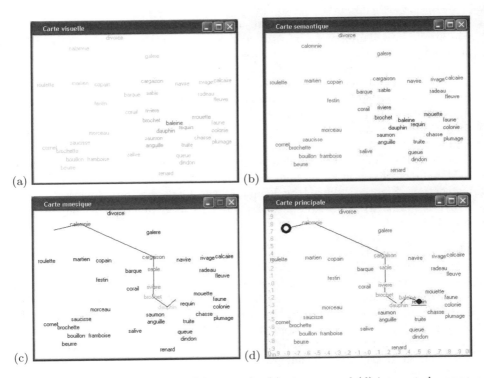

Fig. 2. Examples of (a) visual, (b) semantic, (c) memory and (d) integrated maps

Let us now detail how we implemented each map. Each one dynamically assign a weight to each *word*, conditional on the current scanpath. These weights are in-between 0 and 1: the higher the more attractive.

Visual map. Two kinds of information could affect bottom-up gaze movements: the physiological features of the retina which tend to promote local areas over further ones, and the visual saliency of the current image. Our visual map is therefore a classical visual saliency map multiplied by a filter corresponding to physiological parameters, namely the visual acuity per degree of eccentricity.

$$weight_v(word) = visualSaliency(word) * visualAcuity(word, currentFixation)$$

In our experiment, the saliency of each word only depends on the printed surface of the word, roughly computed as the square of its font size multiplied by its number of characters and normalized to be in the [0,1] range.

$$visualSaliency(word) = nbChar(word)/9 * (fontSize(word)/19)^2$$

The visualAcuity function depends on the distance between the current fixation and the given word, following the classical curve of visual acuity as a function of eccentricity.

Semantic map. As we mentioned previously, the underlying assumption of our semantic component is that spatial proximity reflects semantic similarity: things that belong to the same category are usually near each other in our world. If you are seeking cauliflowers in a brand new supermarket and you are in front of laundry soaps, you would better avoid the current area and search elsewhere. However, if you are in front of carrots, you are almost there! We implemented this idea in the following way: first we computed the semantic similarity between the fixated word and the instruction, using LSA. If this similarity is under 0.2[1], the weights of the current zone are given low values, following a Gaussian around the current fixation. If the similarity is above 0.2, the weights are given higher values still following a Gaussian around the current fixation. The height of the Gaussians depends on the similarities: weights are maximum for a word close to the current fixation and a high similarity between the current word and the instruction. Figure 3 shows the height of the Gaussian as a function of the semantic similarity and examples of Gaussians as a function of the distance from the current fixation.

Memory map. Humans generally do not move back to locations previously visited, although this can be sometimes observed. In the literature, this functionality is usually implemented by means of an inhibition of return mechanism which prevent models from moving back to the previous fixations. In order to have a more flexible mechanism, our memory map strongly decreases the weight corresponding to the last fixations, but slightly increases all previously visited fixations. The latter aims at modeling a forgetting mechanism, so that the model could still go back to words that were visited several saccades before. Basically,

---

[1] This value of 0.2 is usually considered in the LSA literature as a threshold under which items are unrelated.

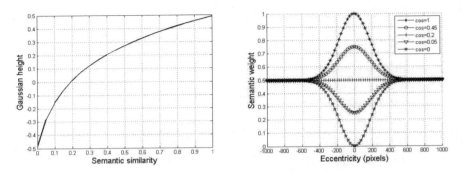

Fig. 3. (a) Semantic Gaussian height as a function of semantic similarity; (b) Examples of Gaussians as a function of the distance from the current fixation.

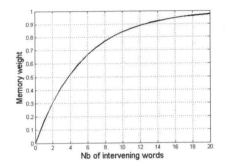

Fig. 4. Memory weights as a function of the number of intervening words

the current fixation is given a weight of 0, meaning that we do not want to go back to it, and weights increase as the number of intervening words in the scanpath increases (Fig. 4).

Map integration. Our model was designed in order to limit as much as possible the number of free parameters. We tried to put a cognitively-based rationale behind each component parameters. However, we did not set the weight of each component in the general performance of the model. Which component is playing the larger role? Is the semantic process important in such a task? Is memory so crucial? Is visual information necessary at all? There is no reason to assume that visual, semantic and memory processes play identical roles. Therefore, they are integrated to form the general map by means of a weighted sum:

$$M_{general} = \alpha_V.M_{visual} + \alpha_S.M_{semantic} + \alpha_M.M_{memory}, \qquad \alpha_V + \alpha_S + \alpha_M = 1$$

The comparison to experimental data will tell us which are the best values for these weights. Before explaining this comparison, we now present the way experimental data were collected.

4   Comparison to Experimental Data

4.1   Experiment

Participants. Forty-three students (average age 20.9 years) participated in the experiment. Each received 5 € for their participation in this experiment, and all had normal or corrected-to-normal vision.

Stimuli. Images were displayed on a computer monitor (1024*768 pixels) at a distance of 50 cm from the seated participant, generating a image subtending 42 horizontal deg. of visual angle.

Procedure. Each subject was in the semantic condition or not and there were 6 images with visual feature facilitating (closer the target, bigger the font size), 6 neutral, and 6 with randomly visual feature (18 trials). Each trial begin with an instruction, followed by a cross fixation. After the subject gazed their fixation on a corner of the display, an image appeared containing the 40 words and observers were instructed to find the best answer as soon as possible. An image was displayed until the subject responded without maximum delay. None of the images used in the experiments were used during the training.

Apparatus. The right eye was tracked using EyeLink II, which is a head mounted eye tracking device (500 Hz Sampling Rate). This was used to record the position and duration of fixations during the search.

4.2   Comparison

Remind that our goal is to find the α_V, α_S and α_M parameters. Our method is to compare human and model scanpaths on various combinations of α_V, α_S and α_M in order to highlight relevant values. Directly comparing human scanpaths and model scanpaths would not be informative though, because there is too much variability in the way a target is found. Therefore, we had to select high-level variables which would be able to characterize such scanpaths.

Variables. We defined the following variables:

- **number of fixations until target is found.** This variable is an indication of the difficulty participants had in finding the target;
- **average angle between saccades.** Humans tend to rationalize their scanpaths in order to minimize their effort and this variable is able to characterize the general shape of the scanpath. For instance, humans seldom show 180° angle (half-turn) between saccades.
- **rate of progression saccades.** Saccades could either be progression saccades if the new fixation is closer to the target than was the previous fixation, or regression saccades. This variable aims at characterizing the general behavior of our participants which is to go more or less quickly to the target.

Discriminating power. One manner to investigate the power of these variables is to look at their ability to discriminate humans scanpaths with respect to different conditions. We compared the average angle between saccades in the semantic and non-semantic conditions. We found a statistical significant difference between these conditions using a Student test (p=.036). In the same way, we compared the rate of progression saccades in the semantic condition only, for the first 6 images the participants saw, for the next 6, and for the last 6. There is indubitably a learning process occuring in the semantic condition of our task because participants should realize after a while that words are semantically organized. We actually found that the rate of progression saccades is able to capture the difference between the first images and the last ones (p<.001). The number of fixations until the target is found is also different in the first 6 images seen and the last 6 (p<.02) Therefore, all variables seem good candidate to compare human and model behaviors.

Fitting. We ran the model for 21 values of α_V and 21 values of α_M both from 0 to 1. α_S is directly obtained since $\alpha_V + \alpha_S + \alpha_M = 1$. To generate several scanpaths per condition, we introduced a bit of noise in the model by not choosing just the best-weighted word in the general map, but randomly selecting between the best one and the second best one. We ran the model 20 times for each combination.

Before going into details in the results, let us discuss some extreme cases.

- If the visual weight α_V is set to a very low value, the model is not dependent on visual information and, more important, is not bound to stay in the local area. The model therefore shows long saccades, sometimes going from one side of the screen to the other.
- If the semantic weight α_S is set to a very low value, the model does not take into account semantic information: it seems to wander from word to word. If the memory is set to a high value, it looks like an exploration behavior.
- If the memory weight α_M is set to a very low value, the model tends to go back to words it has just seen, or even continuously refixates the same word.

The only human data we kept for comparison with the model concerned the semantic condition and, more, the only last 6 images each participant see in this condition, to make sure the semantic organisation was understood by participants. We also removed some bad quality scanpaths. We ended up with a total of 4085 fixations.

We computed the relative error δ of the model data with respect to the human data for each variable and for each combination of α_V, α_S and α_M. Figure 5a displays the relative errors averaged over the three variables for all values of the visual and memory weights. Values corresponding to minimum relative errors for specific variables are marked F (number of fixations), A (angle between saccades) and S (rate of progression saccades). Minimum errors and corresponding weights are the following:

- **number of fixations until target is found:** $\delta = 0.0045$ corresponding to $\alpha_V = 0.3, \alpha_S = 0.1, \alpha_M = 0.6$;

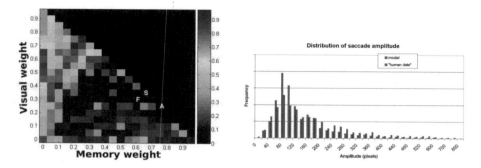

Fig. 5. (a) Average relative errors for all values of α_V and α_M (b) Saccades distribution of human data and best model

- **average angle between saccades**[2]: $\delta = 0.0031$ corresponding to $\alpha_V = 0.25, \alpha_S = 0, \alpha_M = 0.75$;
- **rate of progression saccades:** $\delta = 7.10^{-5}$ corresponding to $\alpha_V = 0.35$, $\alpha_S = 0, \alpha_M = 0.65$;

All three variables give very similar results. Basically, memory plays the highest role. It requires a weight of 60 to 70% to best simulate humans. The visual component needs about 30% and the semantic component plays a minor role in our task.

To perform one more validation, we compared the human distribution of saccades with the ($\alpha_V = .30, \alpha_S = .10, \alpha_M = .60$) model. Actually, a good cognitive model of eye movements should exhibit a distribution of saccades close to the human one. We found a pretty good fit (Fig. 5b) which is another evidence in favor of the respective contributions we found.

5   Conclusion

This paper presents a threefold model of information seeking which was implemented and tested on a word seeking task. We had to make several choices in the design of the 3 components of the model but we tried to be as objective as possible: the visual component is based on the decrease of visual acuity as a function of eccentricity ; the semantic component is based on similarities produced by a cognitive model of semantic associations and the memory component attempts to account for both inhibition of return and forgetting mechanisms. We compared the model output with experimental data and found that the best fit is obtained when the memory is given a strong weight and the semantic component a low weight. This is obviously dependent on our material but we believe that, after being tested on various tasks, this model could be used to predict the

---

[2] We took the second best relative error for this variable since the best one correspond to a non-relevant model with almost no memory and a extremely high level of refixations from which the angle cannot be calculated most of the time.

respective contributions of human visual, semantic and memory processes in a given task. Much remains to be done to apply it to web pages and especially towards more sophisticated visual maps which could be more related to saliency maps. Our next work is to apply this model to images with slightly different visual features but also to images with text paragraphs instead of just words.

References

1. Anderson, J., Matessa, M., Lebiere, C.: ACT-R: A theory of higher level cognition and its relation to visual attention. Human-Comp. Interact. 12(4), 439–462 (1997)
2. Pirolli, P., Fu, W.: SNIF-ACT: A model of information foraging on the world wide web. In: Brusilovsky, P., Corbett, A.T., de Rosis, F. (eds.) UM 2003. LNCS, vol. 2702, pp. 45–53. Springer, Heidelberg (2003)
3. Simon, H.: The structure of ill structured problems. Artificial Intelligence 4(3), 181–201 (1973)
4. Marchionini, G.: Information seeking in electronic environments. Cambridge University Press, Cambridge (1995)
5. Kuhlthau, C.: Learning in digital libraries: an information search process approach. Library Trends 45(4), 708–725 (1997)
6. Guthrie, J.: Locating information in documents: examination of a cognitive model. Reading Research Quarterly 23, 178–199 (1988)
7. Kitajima, M., Blackmon, M., Polson, P.: A comprehension-based model of web navigation and its application to web usability. In: Waern, Y., Cockton, G. (eds.) Proceedings of HCI 2000, People and Computers XIV, pp. 357–373. Springer, Heidelberg (2000)
8. Pirolli, P., Card, S.: Information foraging. Psychological Review 106(4), 643–675 (1999)
9. Fu, W., Gray, W.: Suboptimal tradeoffs in information seeking. Cognitive Psychology 52, 195–242 (2006)
10. Treisman, A., Gelade, G.: A feature-integration theory of attention. Cognitive Psychology 12(1), 97–136 (1980)
11. Itti, L., Koch, C., Niebur, E.: A model of saliency-based visual attention for rapid scene analysis. IEEE Transactions on Pattern Analysis and Machine Intelligence 20(11), 1254–1259 (1998)
12. Mannan, S.K., Ruddock, K.H., Wooding, D.S.: The relationship between the locations of spatial features and those of fixation made during visual examination of briefly presented images. Spatial Vision 10, 165–188 (1996)
13. Yarbus, A.: Eye movements during perception of complex objects. In: Riggs, L.A. (ed.) Eye Movements and Vision, pp. 71–196. Plenum Press, New York (1967)
14. Henderson, J., Brockmole, J., Castelhano, M., Mack, M.: Visual saliency does not account for eye movements during visual search in real-world scenes. In: van Gompel, R., Fischer, M., Murray, W., Hill, R. (eds.) Eye movements: A window on mind and brain, pp. 537–562. Elsevier, Oxford (2007)
15. Navalpakkam, V., Itti, L.: Bottom-up and top-down influences on visual scanpaths. In: Rogowitz, B., Pappas, T.N., Daly, S. (eds.) Proc. SPIE Human Vision and Electronic Imaging XI (HVEI 2006), San Jose, CA, SPIE Press, Bellingham (2006)

16. Feng, G.: Eye movements as time-series random variables: A stochastic model of eye movement control in reading. Cognitive Systems Research 1(1), 70–95 (2006)
17. Landauer, T., Foltz, P., Laham, D.: An introduction to latent semantic analysis. Discourse Processes 25, 259–284 (1998)
18. Klein, R.: Inhibitory tagging system facilitates visual search. Nature 334(6181), 430–431 (1988)
19. Pratt, J., Abrams, A.: Inhibition of return to successively cued spatial locations. Journal of Experimental Psychology: Human Perception and Performance 21(6), 1343–1353 (1995)
20. Tipper, S.P., Weaver, B., Watson, F.L.: Inhibition of return to successively cued spatial locations: A commentary on Pratt and Abrams. Journal of experimental psychology: human, perception and performance 22(5), 1289–1293 (1996)
21. Landauer, T., McNamara, D., Dennis, S., Kintsch, W. (eds.): Handbook of Latent Semantic Analysis. Lawrence Erlbaum Associates, Mahwah (2007)

An Entropy-Based Approach to the Hierarchical Acquisition of Perception-Action Capabilities

David Windridge, Mikhail Shevchenko, and Josef Kittler

Centre for Vision, Speech and Signal Processing,
School of Electronics and Physical Sciences,
University of Surrey, Guildford, UK
Tel.: +44 1483 876043
d.windridge@surrey.ac.uk

Abstract. We detail an approach to the autonomous acquisition of hierarchical perception-action competences in which capabilities are bootstrapped using an information-based saliency measure.

Our principle aim is hence to accelerate learning in embodied autonomous agents by aggregating novel motor capabilities and their corresponding perceptual representations using a subsumption-based strategy. The method seeks to allocate affordance parameterizations according to the current (possibly autonomously-determined) learning goal in a manner that eliminates redundant percept-motor context, thereby obtaining maximal parametric efficiency.

Experimental results within a simulated environment indicate that doing so reduces the complexity of a multistage perception-action learning problem by several orders of magnitude.

Keywords: Perception-Action Architecture, Saliency, Subsumption Hierarchy, Affordance.

1   Introduction

It is by now generally agreed (e.g. [1, 8, 19]) that traditional top-down symbolic approaches to autonomous cognition exhibit significant, and previously unforeseen, complexities. Within such approaches, the underlying mechanism of cognition is assumed to be the manipulation of symbolic representations of the environment via a computational system that utilizes pre-existing syntactic protocols, such as predicate logic [5].

A critical issue for top-down strategies is consequently the implied disparity between the twin processes of symbolic representation and symbolic manipulation. One significant manifestation of this disparity is the problem of *symbol grounding* identified by Harnad [9, 18] in relation to autonomous cognitive agents (i.e. those expected to the exhibit some level of self-determination with regard to their learning processes). Here, because relations between symbols are syntactically constrained via a purely internal mechanism, any connection with the outside world is of a potentially arbitrary and under-determined nature. Mechanisms for addressing this issue have hence typically attempted to constrain

B. Caputo and M. Vincze (Eds.): ICVW 2008, LNCS 5329, pp. 79–92, 2008.

permissible representation via *a priori* sensorimotor linkages imposed at the design level [6, 7, 11, 22].

However, such approaches potentially limit the extent of possible representation, and run contrary to the ideal of cognitive autonomy. They also run the risk of failing to address changes in the environment. Our ideal, in creating a cognitive architecture, ought thus be to create an agent capable of spontaneously generating abstract symbols simultaneously both *representative of*, and *appropriate to*, its surroundings [12], overcoming any potential epistemic circularity implicit in this notion [24].

One such approach is to consider the problem within the terms of a perception-action framework. In Granlund's formulation [8] such an artificial cognition architecture can be seen as constituting the principle that 'actions *precede* perceptions' within the domain of exploration. Autonomous agents exploiting this idea typically attempt to characterize the relationship between their actions and the environmental changes brought about by those actions. In this way initially random exploration of the agent's motor domain can lead to a perceptual model of the environment specified in terms of the *affordances* that it offers the agent [13, 16, 17]. Critical to the success of this approach is the overcoming of the classical artificial cognitive notion that environmental representation needs to be fixed *prior* to the specification of action schemata. A great deal of representational redundancy can thereby by avoided.

In order to generate a cognitive architecture with comparable abstract processing abilities to classical top-down architectures, it is possible to graft a symbol processing system onto the percept-action learner in a manner consistent with this 'representationless' action principle. However, in an ideal system, this symbol processing ability would arise naturally in the context of the perception-action framework.(Various approaches that that can be considered consistent with this idea are set out in [4, 14, 20, 21]).

Our approach to achieving this ideal solution is via the notion of progressive hierarchical perception-action *reparameterization*. In a logical, task-based context such as learning the motor-manipulations required to solve a puzzle, the acquisition of high-level action capabilities (such as the ability to move a puzzle-piece into the solution-state) implicitly characterizes environmental affordance in a symbolic fashion. Thus, if the perceptual representation of the environmental affordance possibilities at the apex of a spontaneously-generated subsumption hierarchy [3] can be efficiently reparameterized so as to remove perceptual context irrelevant to that action capability, then a symbolic representation of the environment is implicitly generated.

The learning strategy outlined in section 2 of this paper thus seeks to acquire behavioral capabilities initially via the unsupervised identification of low-level goals within the agent's *a priori* percept-space, which are then correlated with the action domain via randomized hill-climbing searches in the learning agent's motor space. For the current investigation, these goals are identified via their *information-theoretic* saliency, in contrast to previous purely stochastic approaches [23], such that a large amount of extraneous low-level context is eliminated at the outset (with the result that the final cognitive architecture so constructed will constitute an information-based perception-action hierarchy).

Once acquired, such low-level abilities can then be concatenated and efficiently reparameterized in order to eliminate perceptual invariance in a task-dependent manner, and thereby generate novel perception-action capabilities. Thus, the proposed mechanism can learn via observation of a supervisory agent, parameterizing inferred motor capabilities in terms of salient goals identified, such that replication of *both* the supervisory agent's perceptions and actions becomes possible.

In this way, the perception-action hierarchy encompasses increasingly symbolic manipulation by virtue of the autonomous sub-gaol specification implied in the progressive reparameterization of the subsumption hierarchy so formed. The implicit representation of the environment is thus of hierarchical set of affordances.

Section 3 will thus constitute an experimental examination of the reparameterization methodology within the environment of a shape-sorter puzzle. Results will demonstrate that the information-theoretic approach to perception-action symbolic hierarchy generation is significantly more efficient than non-hierarchical approaches, as well as being resilient to the presence of distractors. Section 4 will conclude by summarizing experimental and theoretical findings.

2   Methodology

2.1   Acquisition of Primitive Percept-Motor Capabilities via Information-Theoretic Saliency

We shall initially define the architecture in generic terms, assuming the existence of an embodied agent capable of undertaking motor actions within the environment. It is further assumed that these actions may be reversed and repeated (so as to permit exploratory and learning behaviors).

A priori motor capabilities in such an agent are defined via the set of independently controllable physical motors $\{C_0, \ldots, C_{K_0}\}$. However, a primitive *behavioral competence* is configured in terms of a vector p that ranges over the agent's *a priori perceptual space*: ie, $C_n = f_n(p)$. Thus, the parameters in terms of which the behavioral competences are defined are not explicitly motor parameters, but rather *perceptual* parameters. The corresponding *a priori perceptual space* is generic in nature, but will usually minimally consist of a topological label space in order to permit labeling of entities that are invariant under translation-like actions (thus enabling basic object-perception). p is hence typically a four-dimensional vector, encompassing the three ordinal directions and a label indexing parameter.

In order to *generate* a primitive behavioral competence it necessary to map the *a priori* motor capabilities onto the perceptual domain. This is achieved within an initially unsupervised learning context via the identification of salient features within the histogram of perceived features (ie the individual components of p) that are generated by randomized exploratory actions. Behavioral competences

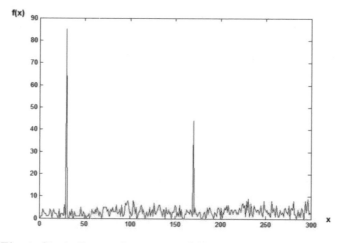

Fig. 1. Single Feature histogram $f(x)$ for 2 Object Environment

can then be indexed via these perceptual goals. Associated with these perceptual goals are a particular set of *perceptual parameters* that are determined by their synchronous behavior: that is, the individual feature peaks are determined to have achieve similar levels of saliency by virtue of exceeding a particular threshold (50% of the maximum). The extraneous components of p are thus eliminated for each goal, and hence much low-level perceptual redundancy can be removed in accordance with the 'action precede perception' principle. Feature *difference* histograms are also calculated to capture behavioral competences that are of an explicitly *relative* character (for instance, the act of aligning one object with another), thus effectively doubling the *a priori* percept-space dimensionality.

Perceptual saliency is determined along the scale-based lines specified in [10], rendering the hierarchical perception-action reparameterization implementation an implicitly information-theoretic one. The scale-saliency approach enables identification of salient features in the perceptual space in a manner resilient to shift/scale changes and to noise; it naturally favors isotropy and geometrical unpredictability, and is consequently suited to extraction of parametric percept elements at the most appropriate scale. We might thus plausibly expect it to favor the characteristics typically attributable to sub-goal and solution states within puzzle environments, such as the identification of key movable objects in a scene, and the identification of the conditions under which these integrate with other geometrically-matching entities. The procedure for obtaining perceptual goals, on obtaining a feature histogram $f(x)$ (such as that given in figure 1 for exploration of a 2D scene with 2 object attractors) is thus to:

(a) Calculate the Shannon entropy $H_D(s, x)$ of local attributes of all points, x of the feature space in question over a range of scales s;

$$H_D(s,x) = -\int p(i,s,x) \ \log \ p(i,s,x) \ di \tag{1}$$

where i is a particular feature-value within the radius s centered on x.
(cf figure 2)

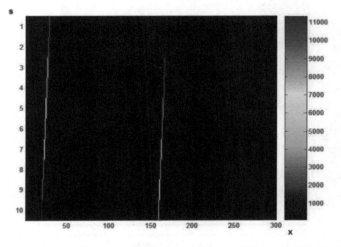

Fig. 2. Shannon Entropy $H_D(s,x)$ at Differing Feature-Space Scales

(b) Select scales at which the entropy over the scale function exhibits a peak, s_p: ie where $H_D(s,x) > threshold$ or where $d[H_D(s,x)]/ds = 0$

(c) Calculate the magnitude change of the PDF as a function of scale at each peak, $W_D(s,x)$:

$$W_D(s,x) = (s^2/[2s-1]) \times \int (d[p(i,s,x)]/ds) \ di \tag{2}$$

(cf figure 3)
This is essentially a measure of scale self-dissimilarity.

(d) The final saliency, S, is then the product $H_D(s,x).W_D(s,x)$ at each peak.
(S is given for the whole scale-feature space in figure 4 to indicate the isolation of peak components)

Suppose, then, that we have obtained a set of perceptual goals that we wish to map onto the motor domain. We select a particular goal g defined by the context-free feature vector: $\boldsymbol{f} = (f_1^g, \dots, f_v^g)$. The distance between the perceptual goal and the current state $\tilde{\boldsymbol{f}}$ is the Euclidean distance; $D(\boldsymbol{f}, \tilde{\boldsymbol{f}}) = [\sum_{i=1}^{v}(f_i^g - \tilde{f}_i)^2]^{\frac{1}{2}}$. Stochastic gradient descent via the method of [15] then enables minimization of this quantity over a number of random instantiations and permutations of the

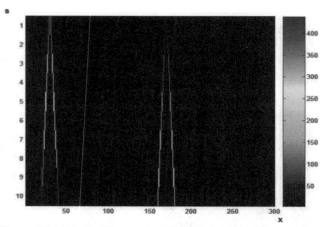

Fig. 3. Magnitude Change in PDF Over Entire Feature Space

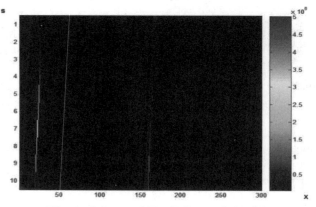

Fig. 4. Saliency S For Differing Sales (s)

motor parameter-space, thereby providing a (partially) context-free mapping between perceptual goals and the agent's action space. (Note that the method [15] finds a global minimum for all components of $\boldsymbol{p}$ simultaneously, rather than considering the motor parameters independently).

Crucially, this method for indexing action capabilities via the perceptual domain can be iteratively generalized to enable construction of the parametric perception-action hierarchy. To do this we consider arbitrary *concatenations* of behavioral competences with an appropriate pruning strategy for non-contributory chains. Thus, if after carrying out the behavioral sequence;

$$C(\boldsymbol{p}) = C_{r_1}(\boldsymbol{p}_{r_1}), C_{r_2}(\boldsymbol{p}_{r_2}), \ldots, C_{r_m}(\boldsymbol{p}_{r_m}) \tag{3}$$

the goal distance $D_i = D(\tilde{\boldsymbol{f}}^{(i)}, \boldsymbol{f}^{(i)})$ does not itself exhibit change for a random parametric instantiation i, then a new sequence is generated. If, on the contrary $D_i \neq D_{i+1}$, then the sequence is deemed relevant, and the gradient descent procedure continued. If $D_i = 0$ at any stage we thus obtain a new behavioral competence $C_{r_{K_0+1}}(\boldsymbol{p}_{r_{K_0+1}})$ where $\boldsymbol{p}_{r_{K_0+1}}$ has a perceptual feature domain given by the tensor product:

$$
\begin{aligned}
\boldsymbol{p}_{r_{K_0+1}} &= \boldsymbol{p}_{r_1} \otimes \boldsymbol{p}_{r_2} \otimes \cdots \otimes \boldsymbol{p}_{r_m} \qquad (4)\\
&= (f_{r_1}^{g_1} \otimes \cdots \otimes f_{r_1}^{g_v}) \otimes \\
&\quad (f_{r_2}^{g_1} \otimes \cdots \otimes f_{r_2}^{g_v}) \otimes \\
&\quad \cdots \\
&\quad (f_{r_m}^{g_1} \otimes \cdots \otimes f_{r_m}^{g_v})
\end{aligned}
$$

Such novel behavioral competences can then in principle be added to existing body of behavioral competences in an iterative manner. The subsumptive nature of the competences so formed implies a hierarchical arrangement of behavioral competences. We will later demonstrate that when such a hierarchy is formed within a supervised puzzle-based environment, this hierarchy naturally reflects the percept-motor sub-goals implicit within the scenario.

Furthermore, in typical operational scenarios, the concatenation $C(\boldsymbol{p})$ has the potential to be efficiently reparameterized (reflecting the fact that goal sub-tasks are not typically independent of each other). It is this property that is key enabling the hierarchical redefinition of the perceptual domain in a manner consistent with the notion of defining the environment in terms of the affordances it offers the active agent.

Parametric Generalization of Behavioral Competences. At the action-level, parametric generalization seeks to make behavioral competences both invariant to environmental configuration and maximally efficient (in the sense of involving no extraneous actions). At the percept level, on the other hand, parametric generalization aims to reorganize the perceptual space associated with behavioral competences in order to define the minimal number of perceptual parameters associated with it. It achieves this by eliminating constant or derived parameter values, and by reindexing multi-dimensional goal parameters into single vectors (if sufficiently few in number).

The former process may be illustrated via an example deriving from the experimental scenario of Section 3: a robotic manipulator-arm equipped with a gripper in a 2D shape-sorter puzzle environment. Suppose, therefore, that there currently exists just *three* behavioral capabilities; moving the manipulator-arm from any given initial position to (x, y); aligning the manipulator-arm with an object indexed by the parameter n; and the act of closing the gripper on an object within its grasp. These are respectively designated $M(x, y)$, $A(n)$, and $G(s)$ (the latter parameter $s = ['grasp', 'ungrasp']$ encompasses the binary states of the gripper).

A typical (potentially autonomously-derived) sub-task in learning the overall puzzle competence is the acquisition of the ability to move an entity with index

n to a 2D location (x, y). Suppose that a redundancy-free sequence of previous capabilities has already been established that is capable of achieving this:

$$A(n_1), G(s_1), M(x_1, y_1), G(s_2).$$

By the tensor product formulation of Equation 4, this implies a perceptual feature domain ranged-over by the vector: $(n_1, s_1, x_1, y_1, s_2)$. However, it is apparent that only three of these features, (n, x, y), function as parametric variables within the act of moving an entity n to the location (x, y); the variables s_1 and s_2 are always set to *constant* values; $s_1 = 'grasp'$, $s_2 = 'ungrasp'$. It is consequently not necessary to (externally) designate them as parameters within the behavioral competence 'moving an entity n to the location (x, y)'. Such variable constancy can always be straight-forwardly determined via sequential random instantiation of parameters.

We are thus able to remap the perceptual space $\boldsymbol{p_{r_{K_0+1}}}$ of novel behavioral competence $C_{r_{K_0+1}}(\boldsymbol{p_{r_{K_0+1}}})$ (ie 'place object n at (x, y)') onto a parametrically smaller perceptual domain $\boldsymbol{p_{r_{K_0+1}}} \rightarrow \boldsymbol{p'_{r_{K_0+1}}}$, where $|\boldsymbol{p_{r_{K_0+1}}}| = 5$ and $|\boldsymbol{p'_{r_{K_0+1}}}| = 3$. Furthermore, if there is any redundancy within the randomly-generated sequence (for instance, if we had obtained an inefficient, but goal-equivalent sequence;

$$G(s_1), G(s_2), A(n_1), G(s_3), M(x_1, y_1), G(s_4),$$

with spurious initial grasping movements $G(s_1), G(s_2)$), then a similar reduction may be achieved by randomly instantiating random parameter subsets and removing any unnecessary ones. Equally, this procedure can establish whether there exist *functionally identical* variables (such as, for instance, when a spatial parameter X_1 requires the same input as a second, apparently differing, instantiation of a singular spatial variable; X_2). Thus, the previous random instantiation procedure enables dimensionality reducing projections of the form $(X_1, X_2) \rightarrow X_1$ (although only for variables of the same type - eg spatial ordinates).

The second major procedure for reduction of the parameter space dimensionality of acquired behavioral competences involves establishing whether concatenated feature-spaces produce a feature histogram in which only a subset of features grow synchronously during trials (measured via the indicated scale-saliency formulation). In this case, the parameter space can be appropriately reduced. An example of this occurs in the experimental scenario of section 3 when the competence 'put object into hole' is learned. In this case, the apparent parameter space, (n, x, y), (consisting in an index n and a hole-location (x, y)) can in fact be reduced to a single parameter n because of the unique (but not *a priori* deducible) correspondence between objects and holes established via exploratory trails, or via observing the supervisor. Similarly, the final, top-level competence 'solve the puzzle' is concatenated from all the instantiations of the sub-competence 'put object n into corresponding hole'. However, the top-level competence is *order independent*, and the final state (being identical for any initial configuration) does not depend upon the object labels; in fact this has no parameters at all.

3   Experimental Results

We aim, in the following experimental test, to ascertain the relative performance of an agent implementing hierarchical information-theoretic perception-action learning in relation to that of a non-hierarchical learner.

The experimental scenario consists in a computer-simulated robotic manipulator-arm equipped with a gripper capable of manipulating a 2D shape-sorter puzzle environment. In this environment, there exist four shapes with corresponding holes into which they may be placed if correctly orientated. The four shapes are otherwise free to occupy the 2D surface, and can be stacked on top of each other. The shape centroids and orientations are randomly in-stantiated at the start of each game (ie following successful completion of the preceding game).

Solution states for the puzzle are characterized by all pieces being in their corresponding holes. The independent *a priori* motor capabilities available to the agent are: Translation of the gripping arm to an arbitrary location (x, y); Grasping and ungrasping of the gripper; Orientation of entities within the gripper to a given absolute value with respect to some fixed reference. The first and last capabilities' motor parameter spaces are continuous; the gripper grasping space is binary in nature. (The binary grasping mechanism might correspond to an automated fine-control perception-action sub-process within a real-word implementation). The *a priori* motor parameter-space is consequently of scope $s \otimes x \otimes y \otimes \theta$, where $s \in \{0, 1\}$, $x \in [0, 1)$, $y \in [0, 1)$ and $\theta \in [0, 2\pi]$.

The *a priori* perceptual parameter-space ranges over $s \otimes x \otimes y \otimes \theta \otimes n \otimes h$ where x, y and θ are the spatial and orientational ordinates associated with the entity n of type h. n is thus an index that differentiates entities (presumably distinguished via sets of individual characteristics such as shape, texture or color in a real-world implementation) of type h. h is thus a coarser-grained class attribution generated on the basis of the previous characteristics via either supervised or unsupervised pattern recognition. In the simulated environment, we assume that entities are separated into pieces and holes on the basis of these characteristics (perhaps by the differing shading characteristics of raised and sunken entities, respectively). However, it is not yet the case that the distinction between pieces and holes has any semantic content; this is what we aim to accomplish with the entropy-based hierarchical perception-action reparameterization.

Initial determination of the primitive behavioral goals is accomplished via the scale-saliency algorithm, followed by gradient descent in the motor parameter space. Distractor entities are included in the perceptual domain (ie they have an index n and corresponding x, y and θ values), but which cannot be moved be the gripper arm (ie they are 'glued' to the table). In order to permit maximum generalizability, these are assumed to be visibly distinguishable from pieces and holes, and so have a different type allocation, h_d.

A human supervisor is observed solving the puzzle over a large number $O(10^3)$ of trials conducted via a 'drag and drop' mouse interface. The solution involves four principle stages of motor-competence (given the initial motor parameter space), which are given an explicit emphasis during training, by emphasizing

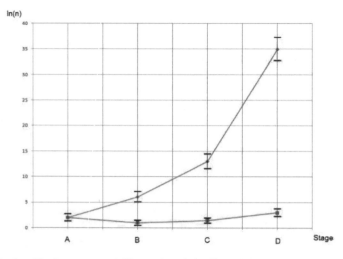

Fig. 5. Relative Performance of Hierarchical (red) and Non-hierarchical (blue) P-A Learning Agents (n = number of P-A cycles)

their sequentially-applied nature. (i.e., the supervisor is instructed not to perform continuous composite movements during training, much as when training a human child). Other than this, no *a priori* knowledge is given to the system. The four key stages of competence are; Moving the gripper to, and aligning it with a given object; Moving a given object to a specified location, Inserting a given object into the appropriate hole, Solving the shape-sorter puzzle.

It is apparent that these competences form a subsumption hierarchy, with each level critically dependent on the level immediately beneath it. Each of the stages consequently requires a progressively complex concatenation of the primitive motor capabilities. Under normal circumstances, it would therefore be the case that the hierarchy of competences would have associated with it a progressively complex perceptual domain (at the first stage, the system observes *oriented entities*; at the second-stage, the system observes *movable objects*; at the third stage the system observes *puzzle pieces*; at the final stage the system simply observes a *puzzle*). Without a modification of the perceptual parameter spaces, the key perceptual entities at each stage would require highly complex description in terms of the *a priori* parameters.

The saliency-based perceptual-action hierarchy formation mechanism should therefore be able to identify these stages of behavioral competence autonomously, and make the appropriate modification to the perceptual space at each level of the hierarchy. In doing so, we expect that it will significantly reduce the parameter space associated with exploratory moves (which are defined at the apex of the hierarchy, and transmitted down the perception-action hierarchy, acquiring increasing amounts of perceptual context as the sub-goals are progressively defined).

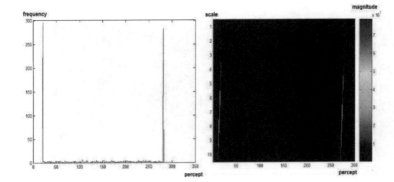

Fig. 6. Single Feature Perceptual Goals for the Competence 'moving from A to B'

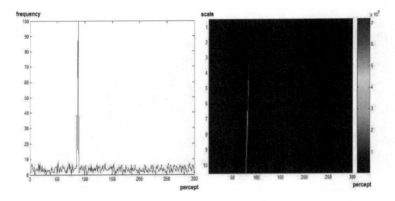

Fig. 7. Single Feature Perceptual Goals for the Competence 'Aligning with an Object'

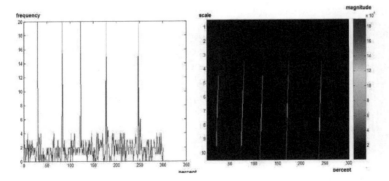

Fig. 8. Single Feature Perceptual Goals for the Competence 'Moving an Object' (5 Objects in Scene)

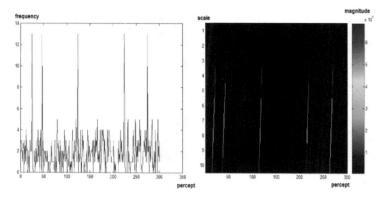

Fig. 9. Single Feature Perceptual Goals for the Competence 'Filling a Hole' (5 Holes in Scene)

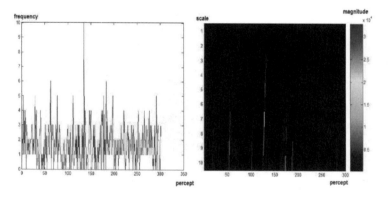

Fig. 10. Single Feature Perceptual Goals for the Competence 'Solving the Puzzle'

To demonstrate this, we establish the total number of perception-action cycles required to achieve the various stages of competence for both a hierarchical learner, and a similar saliency-based perception-action learner, albeit without perceptual reparameterization capability. These results are outlined in figure 5.

We also indicate in figures 6 to 20 typical perceptual goals identified by the scale-saliency mechanism for a single spatial feature at each stage of the hierarchy. In each case, it is apparent that well-defined and relevant perceptual parameters are isolated by the algorithm.

4   Conclusions

We have presented a novel scale-saliency based technique for building perception-action hierarchies. The mechanism is contiguous over the entire hierarchy, and aims to identify both salient parameter sets and parameter indices within the perceptual domain. This enables an affordance-based characterization of the environment in which higher levels of the perception-action hierarchy represent increasingly task-specific and de-contextualized sensorimotor competences. Equivalently, this constitutes a mechanism for autonomous nesting and scheduling of sub-goals; in essence a subsumption hierarchy in the manner of Brooks [2], although constructed spontaneously according to the task requirements.

It is evident from figure 5 that the autonomous generation of an entropy-based saliency hierarchy has the further advantage of significantly more efficient learning than non-reparameterizing methodologies. In fact, it exhibits an essentially linear scaling in performance over the task-complexity hierarchy (similar to an earlier stochastic approach to autonomous construction of perception-action hierarchies [23]). Here, however, the perceptual goals are far more clearly defined for both the primitive and derived capabilities. The method is also significantly resilient to the presence of distractors.

Acknowledgement. The research leading to these results has received funding from the European Community's Seventh Framework Programme (FP7/2007-2013) under grant agreement n° 215078.

References

1. Brooks, R.A.: Intelligence without representation. Artificial Intelligence 47, 139–159 (1991)
2. Brooks, R.A.: New approaches to robotics. Science 253, 1227–1232 (1991)
3. Brooks, R.A.: A robust layered control system for a mobile robot, pp. 2–27 (1990)
4. Castellano, G., Fanelli, A., Mencar, C.: A neuro-fuzzy network to generate human-understandable knowledge from data. Cognitive Systems Research 3, 125–144 (2002)
5. Fodor, J., Pylyshyn, Z.: Connectionism and cognitive architecture: A critical analysis. Cognition 28, 3–71 (1988)
6. Giles, L., Omlin, C.W.: Extraction, insertion and refinement of symbolic rules in dynamically driven recurrent neural networks. Connection Science 5, 307–337 (1993)
7. Kraetzschmar, G.K., Sablatnoeg, S., Enderle, S., Palm, G.: Application of neurosymbolic integration for environment modelling in mobile robots. Hybrid Neural Systems (2000)
8. Granlund, G.: Organization of architectures for cognitive vision systems. In: Proceedings of Workshop on Cognitive Vision, Schloss Dagstuhl, Germany (October 2003)
9. Harnad, S.: The symbol grounding problem. Physica D 42, 335–346 (1990)
10. Kadir, T., Brady, M.: Scale saliency: A novel approach to salient feature and scale selection (2003)
11. Kaelbling, L.P.: Hierarchical learning in stochastic domains: Preliminary results. In: International Conference on Machine Learning, pp. 167–173 (1993)
12. Markman, A., Dietrich, E.: Extending the classical view of representation. Trends Cognit. Sci. 4, 470–475 (1991)
13. McGrenere, J., Ho, W.: Affordances: Clarifying and evolving a concept. In: Proceedings of Graphics Interface 2000, Montreal, Canada, pp. 179–186 (2000)

14. Sharkey, N., Ziemke, T.: Mechanistic vs. phenomenal embodiment: Can robot embodiment lead to Strong AI? Cognitive Systems Research 2(4), 251–262 (2001)
15. Spall, J.C.: Introduction to Stochastic Search and Optimization. John Wiley & Sons, Inc., New York (2003)
16. Steedman, M.: Formalizing affordance. In: 24th Annual Meeting of the Cognitive Science Society, Fairfax VA, pp. 834–839 (2002)
17. Stoytchev, A.: Toward learning the binding affordances of objects: A behavior-grounded approach. In: Proceedings of AAAI Symposium on Developmental Robotics, Stanford University, March 21-23, 2005, pp. 17–22 (2005)
18. Sun, R.: Symbol grounding: a new look at an old idea. Philosophical Psychology 13, 149–172 (2000)
19. Sun, R., Merrill, E., Peterson, T.: From implicit skills to explicit knowledge: a bottom-up model of skill learning. Cognitive Science 25(2), 203–244 (2001)
20. Tano, S., Futamura, D., Uemura, Y.: Efficient learning by symbol emergence in multilayer network and agent collaboration. In: The Ninth IEEE International Conference on FUZZ IEEE 2000, pp. 1056–1061 (2000)
21. Ueno, A., Takeda, H., Nishida, T.: Learning of the way of abstraction in real robots. In: Proceedings of IEEE International Conference on IEEE SMC 1999, vol. 2, pp. 746–751 (1999)
22. Wermter, S.: Knowledge extraction from transducer neural networks. Applied Intelligence: The International Journal of Artificial Intelligence, Neural Networks and Complex Problem-Solving Techniques 12, 27–42 (2000)
23. Windridge, D., Kittler, J.: Open-ended inference of relational representations in the cospal perception-action architecture. In: The 5th International Conference on Computer Vision Systems (ICVS 2007), Bielefeld, March 21–24 (2007), http://biecoll.ub.uni-bielefeld.de/volltexte/2007/89, doi:10.2390/biecoll-icvs2007-172
24. Windridge, D., Kittler, J.: Epistemic constraints on autonomous symbolic representation in natural and artificial agents. In: Smolinski, T.G., Milanova, M.G., Hassanien, A.E. (eds.) Applications of Computational Intelligence in Biology: Current Trends and Open Problems. Studies in Computational Intelligence (SCI), vol. 1. Springer, Heidelberg (2008)

The Role of Implicit Context Information in Guiding Visual-Spatial Attention

Andrea Schankin[1], Olaf Stursberg[2], and Anna Schubö[1]

[1] Ludwig Maximilian University Munich, Department Psychology, Munich, Germany
schankin@psy.lmu.de, anna.schuboe@lmu.de
[2] Technical University of Munich, Institute of Automatic Control, Munich, Germany
stursberg@tum.de

Abstract. Flexibility and adaptability are desirable features of cognitive technical systems. However, in comparison to humans, the development of these features for technical systems is still at the beginning. One approach to improve their realization is to study human cognitive processes and to develop appropriate algorithms, which can be transferred and implemented into technical systems. One example for a typical task common to humans and robots is to find a specific task-relevant object (or target) among other similar but task-irrelevant objects (or distractors). Although this task is quite demanding, humans are doing well in finding task-relevant objects even in unknown environments by applying specific search strategies. For example, when an object is located in a familiar rather than in a new context, humans use the context information to localize the object without recognizing the context as familiar. This phenomenon is known as *contextual cueing*: it is supposed that implicitly learned context information of the environment, i.e. the spatial layout of objects and their relations, guides visual-spatial attention to the target location and thus helps to localize the task-relevant object.

However, in most of the previous psychological studies, artificial objects were used to investigate this effect and thus the ecologic validity is at least dubious. Therefore, the study reported here uses natural objects (LEGO® bricks). In contrast to artificial objects, natural objects are not only different in their visual features but also in their action relevance. Visual search is found to be faster and more accurate when the target is presented within a familiar context and when the knowledge about the context is implicit. This result is encouraging for further adaptation of stimulus material as well as for transferring psychological knowledge to technical applications.

Keywords: Cognitive systems, contextual cueing, learning, perception, visual attention.

1   Introduction

Up to date, intelligent behavior of cognitive technical systems is restricted to functions of rather moderate complexity and to contexts which do not significantly differ from those considered explicitly during the design of the planning

B. Caputo and M. Vincze (Eds.): ICVW 2008, LNCS 5329, pp. 93–106, 2008.

components of the system. Cognitive systems need the ability to reason, plan, solve problems, think abstractly, comprehend complex ideas, and to learn quickly from experience in addition to perceiving their environments and carrying out actions. While cognitive technical systems are still limited in many respects, human visual processing is extremely powerful. For example, humans can recognize an object or an animal within complex scenes with a single glimpse [1]. Thus, one logically reasonable approach to improve the adaptability of technical systems is to study human behavior. Investigating human cognitive processes, modeling and transferring theses mechanisms to technical systems may improve their ability to adapt their behavior to changing environments.

One common but also difficult visual task is to find a task-relevant object (or target) among other similar but task-irrelevant objects (or distractors), e.g. looking for a friend in a crowd or trying to find a car on a parking lot. The human visual system has limited capacity, i.e. not all complex objects can be processed simultaneously. Thus, the system has to select information for more detailed processing by the mechanism known as *selective attention*. The human visual system developed various strategies to perform search tasks quite efficient. On the one side, the physical properties of the target object and their dissimilarity to other objects in the scene provide relevant information for target selection. These so-called bottom-up mechanisms, such as local feature contrast [2, 3], or abrupt onsets [4] guide the deployment of attention. But also top-down, knowledge-based factors, such as novelty [5], familiarity [6] and expectancy [7, 8], influence the guidance of attention. Knowledge on a scene and the relevant information in it can either be given explicitly, e.g. via instruction, or it may be acquired implicitly over time. While explicit mechanisms require additional resources – i.e. the instruction has to be encoded, actively stored and retrieved from working memory – implicit learning mechanisms allow the visual system to quickly extract stimulus regularities [9]. Examples are the repetition of a target's color or location [10, 11], a repeated sequence of target locations [12], or the set of other objects within which the target is presented [13]. The implicit learning requires almost no additional costs, i.e. it does not recur on limited memory resources.

Psychological research has intensively investigated the role of attention in selecting visual information. One main issue is that most of these psychological experiments were performed under laboratory conditions using mostly artificial stimulus material. Recent experiments, however, showed that humans may even perform better when more natural stimuli are used, since far less attention is needed when observers have to process natural scenes rather than artificial stimulus material [14]. In real-world, objects occur in a relatively constant spatial relationship to each other. A stable, meaningful scene structure may thus be used to help guide visual attention to behaviorally relevant targets and may serve to constrain visual processing. Global properties of an image can prioritize objects and regions in complex scenes for selection, recognition, and control of action. This effect is known as *contextual cueing*. In a series of studies Chun and Jiang [13, 15, 16] have demonstrated that the search for a target object was faster when it was embedded in an invariant configuration.

2 Contextual Cueing Experiments with Natural Stimuli

In contextual cueing experiments, participants search through a display of objects in order to find and identify a target object. The visual context, as previously described as the spatial relationship between objects, is defined by the spatial arrangement of the distractor objects. In order to investigate the effects of context, some of the search displays are repeated across blocks throughout the entire experimental session. Because the relationship between the objects is held constant, the context predicts the target location (but not the identity of the target). In a respective control condition, new displays, which are randomly generated for every trial in each block, are presented intermingled with old context displays. Although reaction times (RTs) decrease for both old and new contexts during the time course, there is an additional learning effect that differs for old and new contexts. After four to five repetitions, RTs in the repeated context conditions become faster than RTs in the newly generated context conditions, even though participants were not told explicitly that the global context structure was informative. Probably, an association between the spatial arrangement of the distractors (or context elements) and the target location was formed and is subsequently used to guide visual-spatial attention to the target location (context learning). Interestingly, contextual knowledge is acquired through implicit learning processes. In contrast to *explicit learning*, which is characterized as an active process where people seek out the structure of any information that is presented to them, *implicit learning* is a passive process by which complex information about the stimulus environment is acquired without intention (i.e., learning occurs incidentally, simply by the exposure to the information) or awareness [15].

It is important to note here that learning (how knowledge is acquired) and memory (how knowledge is stored) are distinct processes [17]. Thus, implicit learning can produce explicit knowledge, and certain forms of explicitly learned information are only accessible through implicit measures. Which kind of knowledge (implicit or explicit) is acquired can be tested by so-called implicit and explicit tests of memory. For example, after a learning phase, participants are presented with a list of previously seen and new objects and asked to categorize them as old or new (recognition test). This is a direct test of which objects have been learned (*explicit knowledge*). Implicit measures, on the other side, do not require the participants to remember any information. The effect of learning is expressed by a faster and/or more accurate response to old relative to new objects (*implicit knowledge*).

In typical contextual cueing experiments, the context information is implicit. That is, participants are usually not aware of the fact that some of the contexts were repeatedly shown. This was measured by a so-called forced-choice recognition test, in which participants had to categorize old and new displays as already seen or new. If the knowledge about previously seen contexts is implicit, participants are not able to distinguish between both display types; what is basically the findings in such procedures.

Investigating how this implicit memory-based mechanism works may allow to implement similar principles in cognitive technical systems for improve their adaptivity to changing contexts. However, existing results on implicit context representations were obtained by using rather artificial stimulus material, e.g. the rotated letters T and L [15, 18], while cognitive technical systems should act in real-world scenarios. It is difficult to predict whether a simple transfer of the results would be possible, i.e. additional experiments were required to test whether context information can be used to guide attention also in real-world scenes.

To adapt the artificial contextual cueing paradigm to real-world scenarios, there are three possible strategies: the context, the objects, or both could be changed from artificial to natural. So far, only a few studies have addressed this issue [19, 20]. In these studies either only the context was changed (a natural scene, in which a T or L had to be found [20]) or also the target object was replaced by a natural object [19]. Although the contextual cueing effect was still present, the former implicit knowledge about the association of context and target position found in previous experiments became explicit. This means that the participants noticed the repetition of some displays and were able to correctly categorize displays as being old or new. This switch from implicit to explicit knowledge may be due to a better recognition of natural scenes in contrast to artificial object arrangements. However, also a decreased number of scenes used in these experiments may account for the effect, as the reduction in scenes may have increased the probability of recognition. In any case, although the knowledge became explicit, the learning was still incidental: participant's were neither instructed to encode the context nor were they told that the context would provide any relevant information.

The goal of the experiments reported in this investigation was to design a paradigm, which allows to replicate the contextual cueing effect on the one side and to keep the learning process implicit on the other side. Therefore, the 'classical' paradigm, in which context is defined by the layout of distractors, is modified by replacing artificial line-object context elements by natural objects. Natural objects differ from artificial objects not only by their visual features, such as the spectrum of colors, contours, or luminance. More important, perceiving natural objects also activates object-specific action goals [21, 22, 23]. For example, perceiving a tool, like a hammer, also activates possible actions, like using it to drive a nail into the wall. That is, natural objects probably require more cognitive capacity to be processed. Besides the perceptual analysis, additional cognitive processes (e.g., forming an intention for action or action planing) have to take place. Thus, asking whether contextual cueing effects are affected by the stimulus type (articial vs. natural) is not trivial. Although it is known that implicit learning needs less attentional resources than explicit learning, additional cognitive processes may reduce implicit learning speed.

The approach of using natural objects, as presented here, is also interesting for application to technical systems. For example, robots are often required to find a task-relevant object among task-irrelevant objects, such as specific tools or components. In this case not the whole background provides context information, as was investigated by the real-world studies cited above, but rather the arrangement of tools and components.

3   Methods

3.1   Participants

Sixteen paid volunteers (14 women, 2 men, aged between 19 and 29 years, mean age 22.6 years) participated in the experiment. Three participants were left handed, all others right handed. All participants reported normal or corrected-to-normal visual acuity. The study was carried out in accordance with the ethical standards laid down in the 1964 Declaration of Helsinki.

3.2   Stimuli and Apparatus

Participants were seated in a comfortable armchair in a dimly lit and sound attenuated room with response buttons located under their left and right index fingers. All stimuli were presented on a 17 inch computer screen placed 100 cm in front of the participants at the center of their field of vision. Participants were instructed to search for one of two task-relevant objects (the target) among other task-irrelevant objects (the distractors) and identify the target by pressing a response key.

In Fig. 1, a sample search array (left) and an enlarged object (right) are shown. Each search array consisted of 12 LEGO® objects (1.2 ° in visual angle), which could appear within an invisible matrix of 12 x 9 locations that subtended

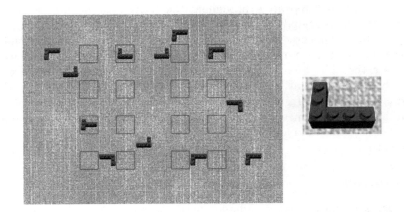

Fig. 1. Example of a search array (left). Each object was constructed by 2 LEGO® bricks (right). The task-relevant object (or target) was a rotated T, the task-irrelevant objects (or distractors) were rotated L objects. All objects were presented in red on a light blue background. The target could appear at one of the marked locations (not visible in the actual experiment).

approximately 15.1 x 10.2 ° in visual angle. The objects were designed by a CAD program (LegoCad). The software, which was developed by Lego in cooperation with Autodesk, allows constructing simple 3D models and machines. Two red LEGO® bricks were used to build one object. To make the stimulus set more realistic they were additionally rotated in perspective so that they appeared to lie on a surface. Using these objects was a compromise between constructing natural objects on the one side and keeping the objects comparable to the artificial objects, which were usually used in contextual cueing experiments on the other side. Thus, although our objects were still similar to previous experiments, which presented a rotated T among a set of rotated L, they were also action relevant. In contrast to abstract letters, LEGO® bricks can be grasped, moved around and used to assembly a more complex object.

In the experiment, the target was a LEGO® object in the form of a T, rotated 90 ° to the right or left. The target was presented at one of 16 selected locations of the 12 x 9 matrix, and the target positions were identical for all participants (cf. Fig. 1). Each of these 16 locations was used once in old and once in new configurations. The task-irrelevant objects (or distracting context elements) were 11 L-shaped LEGO® objects, which were presented randomly in one of four rotations (0 °, 90 °, 180 °, 270 °). The distractor locations in each configuration were randomly sampled from all 108 possible locations, including target locations used in other configurations. In each configuration, half of the objects were placed left and the other half right of fixation, balanced for eccentricity. Configurations were generated separately for each participant.

Similar to previous experiments, the visual context was defined as the arrangement of distractor objects. In order to measure the influence of context, old and new configurations were presented throughout the experiment. The old set of stimuli consisted of 16 configurations, randomly generated at the beginning of the experiment then repeated throughout the entire experimental session once per block. In the repeated configuration condition, the target (left- or rightward oriented) always appeared in the same location within any particular configuration and the identities of the distractors within their respective spatial locations were preserved. The target type (left- or rightward pointing T), however, was randomly chosen so that the identity of the target did not correlate with any of the configurations. In contrast, in the new set of stimuli configurations of distracting context elements were generated randomly on each trial. The differences of RT between old and new configurations can be interpreted as the effect of contextual learning.

3.3   Procedure

Participants performed three different parts: training at the beginning of the experiment, followed by the actual experiment, and a recognition test at the end. Participants were instructed to search for a rotated T and press one of two buttons corresponding to whether the bottom of the T pointed to the right or to the left as soon as they could. They performed three training blocks of 32 trials each. A trial started with a fixation cross appearing in the middle of the screen for 500 msec, followed by the search display that was presented for further 500 msec. Participants pressed a key to indicate the identity of the target (a left- or rightward pointing T). After a brief pause of 1-2 sec, the following trial was initiated by the computer. The training was necessary to familiarize participants with the experimental task and procedure and to minimize inter-subject variability.

The experimental session consisted of 30 blocks of 32 trials each (16 old, 16 new configurations), for a total of 960 trials for each participant. The experimental procedure was exactly the same as in the previous training session. Feedback was given at the end of the block on the percentage of correct responses. Participants were not informed that the spatial configuration of the stimuli in some trials would be repeated, nor were they told to attend to or encode the global array. They were simply given instruction to respond to the target's identity. It was stressed that they were to respond as quickly and as accurately as possible. A mandatory break of about 1-2 minutes was inserted after five blocks each and a longer break after half of the experiment.

At the end of the final block, participants performed a recognition task. All 16 old configurations were presented again, intermingled with 16 newly generated configurations. Participants were asked to classify all configurations as already seen or new, respectively. The recognition served as a control measure: If learning was indeed implicit, participants should not be able to distinguish between old and new displays.

3.4   Data Analysis

Reaction times were measured as the time between onset of the search display and the participant's response. Pressing the wrong button, pressing the button too quickly (<150 msec) and pressing it too slowly (>2000 msec) were defined as errors. To estimate the general learning effect and the time point when context learning occurred, blocks were grouped in sets of 6 blocks each into 5 epochs. Error percentages and mean reaction times, separately for correct and incorrect responses, were entered in separate repeated-measures ANOVAs with factors of context (old vs. new configurations), and epoch (1 to 5). An effect of epoch would reflect changing RTs (or errors) in the time course of the experiment (i.e. general learning), whereas a statistical effect of context would reflect how the repetition of a context affected the search for the target object (i.e. contextual learning). More important, an interaction between both factors would suggest that there was no general difference in visual processing of old and new displays but that the context information was learned over time. In order to demonstrate that the knowledge of display repetition is indeed implicit, a recognition test was performed at the end of the experiment. The hit rate (old displays were correctly categorized as old) was compared to the false alarms rate (new displays were wrongly categorized as old) by a paired t-test. Additionally, the effect of a possible recognition of context repetition on the contextual cueing effect was estimated by a correlation between d-prime as measurement of recognition (according to signal detection theory) and the contextual cueing effect reflected by RT differences between old and new contexts ($\mathrm{RT}_{new} - \mathrm{RT}_{old}$).

4   Experimental Results

Error percentages and reaction time data are presented as a function of epoch and context in Fig. 2. With about 20.9%, error rate was relatively high in this experiment (Fig. 2, left panel). In the course of the experiment, the participants' accuracy increased, as shown by decreasing errors from about 26.8% in epoch 1 to 18.1% in epoch 5 ($F(4, 60) = 17.5$, $p < 0.001$). Planned comparisons revealed a significant decrease from epoch 1 to epoch 2 ($F(1, 15) = 15.9$, $p < 0.01$), and from epoch 2 to epoch 3 ($F(1, 15) = 5.0$, $p < 0.05$). The error rates than remained constant on this level, with $F(1, 15) < 1$. Error rates also varied as a function of context: $F(1, 15) = 11.7$, $p < 0.01$. Searching for a target in an old context was on average more accurate than searching in a new context (22.2% vs. 19.5%). However, a significant interaction between epoch and context ($F(4, 60) = 2.6$, $p < 0.05$), indicates that this advantage developed during the course of the experiment, i.e. an old context had to be learned as being known first. Probably only some of the repeated old displays were learned from block to block, as reflected by a reliable linear trend ($F(1, 15) = 5.5$, $p < 0.05$). Similarly to error rates, the RT of correct responses decreased over time: $F(4, 60) = 6.0$, $p < 0.01$, $\epsilon = 0.561$ (Fig. 2, right panel). Single planned comparison showed a reliable linear trend in search time: $F(1, 15) = 9.0$, $p < 0.01$. The contextual cueing effect, defined as an RT benefit in the old condition compared to the new condition across all epochs, was significant: $F(1, 15) = 19.8$, $p < 0.001$. The significant interaction between epoch and context ($F(4, 60) = 3.8$, $p < 0.01$)

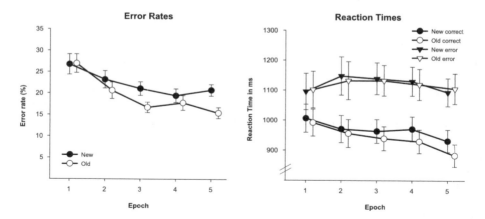

Fig. 2. Error rates (left panel; % error and SEM) and reaction times (right panel; RT and SEM) as a function of epoch (x-axis) and context (filled vs. unfilled symbols). Reaction times were further analyzed separately for correct and incorrect responses (circles vs. triangles).

indicates that performance was similar for both context types at the beginning of the experiment but that learning of the old context led to faster reaction times for the latter when compared to new contexts. The overall contextual cueing benefit, measured as the difference between old and new configurations across the last three epochs (according to [24]) was 38 msec ($SD=29$ msec). In contrast to correct responses, the RT of incorrect responses neither varied as a function of epoch or context nor did both factors interact with each other (for all tests $F < 1$). Similarly to previous studies, the knowledge about repeated configurations was implicit. In the recognition test, the hit rate (42%) did not differ significantly from the false alarms rate (36%): $t(15) = 1.2$, $p = 0.248$. Additionally, d-prime as measurement of recognition did not correlate with the size of the contextual cueing effect ($r = 0.29$, $p = 0.28$), showing that there was no relation between a possible explicit knowledge about the configurations and the size of the contextual cueing effect.

5   Contextual Cueing Effects with Natural Stimuli

The aim of the present study was to investigate whether knowledge acquired with artificial stimuli in laboratory environments can also be obtained by more natural objects and may be transferable to real-world scenarios developed for technical systems. Artificial and natural objects do not only differ in their visual features but also in their action relevance. According to this view, which is different from the classic information processing view, perception and action are not seen as functionally distinguishable stages, but as intimately related processes [23]. Recently, brain studies have shown that perceiving natural objects not only activates visual cortical areas but also motor-related areas, which was not the case for non-manipulable abstract stimuli [25].

Similarly to previous experiments [13, 15, 16], we observed two learning processes in the current experiment. First, generally decreasing RTs in both context conditions reflects a general learning process. And second, finding a target in an old (repeated) context is faster than finding the target in a new context (context learning). The first learning effect is probably due to different underlying cognitive processes, e.g. the familiarization with the task and stimulus material and general training effects in response selection. This interpretation is supported by electrophysiological findings [26].

The second learning effect is often interpreted as guidance of visual attention by the context [13, 15, 16]. In the present type of contextual cueing experiments, observers have to first find the target in the display set of elements and then to identify the target's identity. This task requires the observer to focus her/his attention on the target location. Although the target identity differed in all trials and in all context conditions, the arrangement of the distractors and the target location were kept constant in the old context condition. Observers may thus have used the repetition of the spatial layout in the old context conditions to predict the target location. As this prediction did not include the target identity that was needed for appropriate responding, they still had to allocate their attention to the target location. According to these considerations, faster RTs in the old context condition reflect a faster shift of attention to the target location: or, in other words, the 'familiar' visual context guided visual-spatial attention. Electrophysiological experiments support this interpretation [18, 26, 27].

More interestingly, the results presented here demonstrate that context learning does not only occur when artificial stimuli are used but can also be obtained with natural objects in a very similar manner: Searching for a task-relevant object among task-irrelevant objects was faster and more accurate when the object was presented within a known context. Obviously, the larger visual complexity of natural objects in comparison to simple artificial ones did not influence the learning process per se. Also, probable additional cognitive processes, e.g. intentions or action planing, did not interfer with learning.

In contrast to previous experiments using real-world scenes [19, 20], the knowledge about the context was still implicit in our experiment: participants could not distinguish between already inspected and newly generated displays. One reason might be that in our experiment, similar to artificial stimuli, it was still the arrangement of the context elements that defined the visual context. In this case, the stimulus type (artificial versus natural) does not play a role. If the global context, however, may have a meaning on its own, as is the case in real-world scenes, it seems to be processed in a different way and to get access to explicit memory. The information provided by the arrangement of some few context object elements, however, is probably too abstract to be explicitly accessible from memory. One argument which would favor this interpretation is that learning takes longer in experiments using artificial stimuli, namely about 5 repetitions [15], whereas already the first repetition of a real-world scene produces large contextual cueing effects [20]. Probably, meaningful contexts are easier to distinguish and to categorize, what facilitates learning. In summary, although the global context is a source of complexity and adds to the segmentation problem in vision, contextual cueing experiments show how context also serves to facilitate processing rather than just complicating it. The present findings highlight the important role of implicit learning and memory mechanisms in visual processing.

6   Contextual Cueing in Cognitive Technical Systems

From an engineering perspective, the interesting question is how the principles of contextual cueing and learning can be employed to improve the perceptual and cognitive abilities of autonomous technical systems. The development of, e.g., autonomous robotic systems or ambient intelligence systems involves the real-time analysis of enormous quantities of data. These data have to be processed efficiently to provide "on time availability" of relevant information for solving given tasks. Knowledge has to be applied about what needs to be attended to, and when, and what to do in a meaningful sequence, in correspondence with visual feedback on the actual context situation. Contextual cueing may be one important process that allows shortening the computational time required to estimate where attention should be directed to. For reasons of implementation,

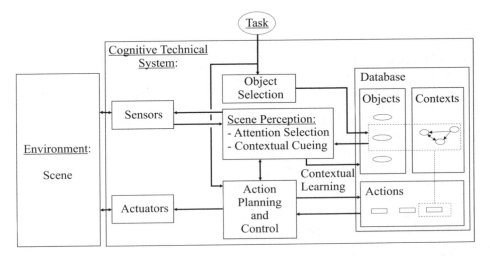

Fig. 3. Architecture of a cognitive technical system for realizing the human principle of attention selection and contextual cueing

it is important to note here that contextual cueing involves at least two different mechanisms: first, an association between the context information of the environment (i.e., the distractor arrangement) and the task-relevant object (i.e., the target) has to be learned ('contextual learning'), and second, this knowledge has to be applied to the current environment to guide attention to the most probable target location ('guidance of attention').

The architecture in Fig. 3 shows a possibility to represent these mechanisms as part of the information processing in a cognitive technical system: Starting point is a task which is given to the system or is autonomously generated based on previous behavior of the environment or the system itself. An algorithm for object selection determines a set of objects known to be relevant for the current task. Object features, together with context information, are taken from a database and are transferred to an algorithm for scene perception – the database plays the role of the human memory for storing explicit and implicit knowledge on situation-dependent context. The perception algorithm analyzes the data on the current environment scene (as received from the sensors) with respect to the selected objects and their embedding in the context obtained from the database. In order to realize contextual cueing according to the principles found for human attention selection, a main future challenge in designing the perception algorithm is to develop new mechanisms for pattern recognition which identify complete configurations consisting of relevant objects **and** contexts. The hope is that such a combined perception leads to a quicker understanding of the momentary environment situation by the cognitive technical system. Newly perceived configurations are stored in the database as object-context pairs, a process which can be understood as contextual learning. The following trade-off has to be resolved for technical realizations of this learning procedure: Obviously, the larger the number of previously experienced object-context pairs is, the higher is the probability of finding a matching configuration in the database for the currently envisaged situation. On the other hand, memory restrictions and particularly the computational effort to find the best-matching object-context pair in the database will slow down the whole perception process once high numbers of experienced configurations are stored. This leads to the requirement of defining similarity measures for contexts, i.e. configurations in which the structure (e.g. the relative positions of objects) differ less than a specified bound are not repeatedly stored in the database. Furthermore, suitable datastructures for fast referencing of contexts have to be developed.

An essential step of perceiving the environment for robotic systems is to control sensors (e.g. camera or radar systems) such that data of environment locations which have high relevance for the present task are obtained first. The block 'action planning and control' in Fig. 3 is understand broader, however, and refers in general to algorithms for computing control signals to be sent to the actuators. This means that information on the current task and on the perceived scene are used by planning algorithms to determine the behavior of the cognitive technical system as required for fulfilling the task. While intensive research on planning algorithms has produced a variety of approaches for different types of systems and tasks (see e.g. [32] for an overview), the challenge is here to directly include the perceived context into the planning and control trajectory. Once computed, the actions (or action sequences respectively) are included in the database with reference to the task, the selected objects, and the context identified to be relevant for the configuration, i.e. the actions are part of the contextual learning procedure. The algorithm for scene perception, contextual learning, and action planning run, of course, concurrently (or iteratively, respectively) until the present task is accomplished. Since most tasks require the consideration of dynamic scenes, sequences of scenes and contexts have to be identified and sequences of actions need to be computed and executed. The challenge for most autonomous technical systems is to apply scene perception, planning, and control in such an inter-connected fashion that real-time response is possible, eventually on the time-scale on which human act. Embedding contextual cueing and learning in integrated algorithms seems to be a possible way of approaching this aim a bit further.

7   Conclusion

Although visual search for a task-relevant object among task-irrelevant objects is a demanding task for the visual information processing, implicit learning mechanisms, such as contextual cueing, allow the visual system to quickly extract stimulus regularities [9]. This mechanism does not only occur with artificial stimulus material but also with natural objects (current study) and real-world scenes [19, 20]. Implementing such a mechanism into cognitive technical systems may help to develop flexible and adaptive behavior.

Acknowledgments

This project was part of the Cluster of Excellence 'Cognition for Technical Systems' (CoTeSys), funded by the Deutsche Forschungsgemeinschaft (German Research Foundation, DFG). We want to thank Christian Stößel for his help in designing the stimuli.

References

[1] Thorpe, S., Fize, D., Marlot, C.: Speed of processing in the human visual system. Nature 381, 520–522 (1996)

[2] Bravo, M.J., Nakayama, K.: The role of attention in different visual-search tasks. Perception & Psychophysics 51, 465–472 (1992)

[3] Treisman, A., Gormican, S.: Feature analysis in early vision: evidence from search asymmetries. Psychological Review 95, 15–48 (1988)

[4] Yantis, S., Jonides, J.: Abrupt visual onsets and selective attention: Evidence from visual search. Journal of Experimental Psychology: Human Perception & Performance 10, 601–621 (1984)

[5] Johnston, W.A., et al.: Attention capture by novel stimuli. Journal of Experimental Psychology: General 119, 397–411 (1990)

[6] Wang, Q., Cavanagh, P., Green, M.: Familiarity and pop-out in visual search. Perception & Psychophysics 56, 495–500 (1994)

[7] Shaw, M.L.: A capacity allocation model for reaction time. Journal of Experimental Psychology: Human Perception & Performance 4, 586–598 (1978)

[8] Shaw, M.L., Shaw, P.: Optimal allocation of cognitive resources to spatial locations. Journal of Experimental Psychology: Human Perception & Performance 3, 201–211 (1977)

[9] Reber, A.S.: Implicit learning and tacit knowledge. Journal of Experimental Psychology: General 118, 219–235 (1989)

[10] Maljkovic, V., Nakayama, K.: Priming of popout: I. Role of features. Memory & Cognition 22, 657–672 (1994)

[11] Maljkovic, V., Nakayama, K.: Priming of popout: II. The role of position. Perception & Psychophysics 58, 977–991 (1996)

[12] Nissen, M.J., Bullemer, P.: Attentional requirements of learning: Evidence from performance measures. Cognitive Psychology 19, 1–32 (1987)

[13] Chun, M.M., Jiang, Y.H.: Top-down attentional guidance based on implicit learning of visual covariation. Psychological Science 10, 360–365 (1999)

[14] Fei-Fei, L., VanRuellen, R., Koch, C., Perona, P.: Why does natural scene categorization require little attention? Exploring attentional requirements for natural and synthetic stimuli. Visual Cognition 12, 893–924 (2005)

[15] Chun, M.M., Jiang, Y.H.: Contextual cueing: Implicit learning and memory of visual context guides spatial attention. Cognitive Psychology 36, 28–71 (1998)

[16] Chun, M.M., Jiang, Y.H.: Implicit, long-term spatial contextual memory. Journal of Experimental Psychology-Learning Memory and Cognition 29, 224–234 (2003)

[17] Buchner, A., Wippich, W.: Differences and commonalities between implicit learning and implicit memory. In: Stadler, M.A., Frensch, P.A. (eds.) Handbook of implicit learning, pp. 3–47. Sage Publications, Thousand Oaks (1998)

[18] Olson, I.R., Chun, M.M., Allison, T.: Contextual guidance of attention: human intracranial event-related potential evidence for feedback modulation in anatomically early temporally late stages of visual processing. Brain 124, 1417–1425 (2001)

[19] Brockmole, J.R., Castelhano, M.S., Henderson, J.M.: Contextual cueing in naturalistic scenes: Global and local contexts. Journal of Experimental Psychology-Learning Memory and Cognition 32, 699–706 (2006)

[20] Brockmole, J.R., Henderson, J.M.: Using real-world scenes as contextual cues for search. Visual Cognition 13, 99–108 (2006)

[21] Greenwald, A.G.: A choice reaction time test of ideomotor theory. Journal of Experimental Psychology 86, 20–25 (1970)

[22] Prinz, W.: Perception and action planning. European Journal of Cognitive Psychology 9, 129–154 (1997)

[23] Hommel, B., et al.: The theory of event coding (TEC): a framework for perception and action planning. Behavioral & Brain Sciences 24, 849–937 (2001)

[24] Kunar, M.A., Flusberg, S.J., Wolfe, J.M.: Contextual cuing by global features. Perception & Psychophysics 68, 1204–1216 (2006)

[25] Grafton, S.T., et al.: Premotor cortex activation during observation and naming of familiar tools. Neuroimage 6, 231–236 (1997)

[26] Schankin, A., Schubö, A.: Cognitive processes facilitated by contextual cueing. Evidence from event-related brain potentials. Psychophysiology (in press)

[27] Johnson, J.S., Woodman, G.F., Braun, E., Luck, S.J.: Implicit memory influences the allocation of attention in visual cortex. Psychonomic Bulletin & Review 14, 834–839 (2007)

[28] Bar, M., Biederman, I.: Subliminal visual priming. Psychological Science 9, 464–469 (1998)

[29] Tulving, E., Schacter, D.L.: Priming and human memory systems. Science 247, 301–306 (1990)

[30] Greene, A.J., et al.: Hippocampal differentiation without recognition: an fMRI analysis of the contextual cueing task. Learning and Memory 14, 548–553 (2007)

[31] Chun, M.M., Phelps, E.A.: Memory deficits for implicit contextual information in amnesic subjects with hippocampal damage. Nature Neuroscience 2, 844–847 (1999)

[32] Ghallab, M., Nau, D., Traverso, P.: Automated Planning. Morgan Kaufmann Publ., San Francisco (2004)

Probabilistic Pose Recovery Using Learned Hierarchical Object Models

Renaud Detry[1], Nicolas Pugeault[2], and Justus Piater[1]

[1] Université de Liège, Liège, Belgium
{Renaud.Detry,Justus.Piater}@ULg.ac.be
[2] University of Southern Denmark, Odense, Denmark,
The University of Edinburgh, Edinburgh, Scotland, UK
npugeaul@inf.ed.ac.uk

Abstract. This paper presents a probabilistic representation for 3D objects, and details the mechanism of inferring the pose of real-world objects from vision. Our object model has the form of a hierarchy of increasingly expressive 3D features, and probabilistically represents 3D relations between these. Features at the bottom of the hierarchy are bound to local perceptions; while we currently only use visual features, our method can in principle incorporate features from diverse modalities within a coherent framework. Model instances are detected using a Nonparametric Belief Propagation algorithm which propagates evidence through the hierarchy to infer globally consistent poses for every feature of the model. Belief updates are managed by an importance-sampling mechanism that is critical for efficient and precise propagation. We conclude with a series of pose estimation experiments on real objects, along with quantitative performance evaluation.

Keywords: Computer vision, 3D object representation, pose estimation, Nonparametric Belief Propagation.

1 Introduction

The merits of part-based and hierarchical approaches to object modelling have often been put forward in the vision community [9,5,11]. Part-based representations are more robust to occlusions and viewpoint changes than global representations, and spatial configurations increase their expressiveness. Moreover, they not only allow for bottom-up inference of object parameters based on features detected in images, but also for top-down inference of image-space appearance based on object parameters.

The advantages of visual part-based representations naturally extend to multi-sensory cases. For example, haptic and proprioceptive information won't relate to an object as a whole. Instead, they typically emerge from specific grasps, on specific parts of the object. Part-based representation offer a neat way to *locally* encode cross-modal descriptions that emphasise the relations between the different types of percepts.

B. Caputo and M. Vincze (Eds.): ICVW 2008, LNCS 5329, pp. 107–120, 2008.

We are currently developing a 3D, part-based object representation framework, along with mechanisms for unsupervised learning and probabilistic inference of the model. Our model combines local appearance and 3D spatial relationships through a hierarchy of increasingly expressive *features*. Features at the bottom of the hierarchy are bound to local visual perceptions. Features at other levels represent combinations of more elementary features, and encode probabilistic relative spatial relationships between their children. The top level of the hierarchy contains a single feature which represents the object.

To detect instances of a model in a cluttered scene, evidence is propagated throughout the hierarchy by probabilistic inference mechanisms, leading to one or more consistent scene interpretations: the model is able to suggest a number of likely *poses* for the object, a pose being composed of a 3D location and a 3D body orientation defined in the reference frame of the camera that captured the raw visual data. The use of probabilistic inference algorithms permits the uniform integration of all available evidence, allowing for unbiased contributions of all low-level features.

In previous work [2], we presented a learning method that constructs a hierarchy from a set of object observations. We also gave an overview of an inference process that followed a straightforward Nonparametric Belief Propagation scheme [12] and allowed for pose recovery of artificial objects. In this paper, we present in greater detail a significantly improved version of this inference process. We added an importance-sampling (IS) message product suggested in a similar form by Ihler et al. [6], and extended it to a two-level IS sampling of *implicit* message products which is readily applicable to pose estimation on real-world objects.

Unsupervised learning, probabilistic representation and robust detection are three aspects that we believe make our representation a good candidate for the perception and memory tasks of a cognitive system. Furthermore, the features organized in the hierarchies are not especially restricted to one input modality. We currently work with visual input only, but our model is intended to unite different types of perceptual information, e.g. vision plus haptic and proprioceptive inputs simultaneously. This will produce cross-modal descriptions and cross-modal behaviors directly applicable to action-related tasks such as grasping and object manipulation, as a grasp strategy may be linked directly to visual features that predict its applicability.

We emphasize that we are not developing an object classification framework. Object classification is best achieved using *discriminative* models and presupposes the presence of one object to be classified. Instead, we intend to develop *object-centric* representations that allow for detection and localisation of known objects within a highly cluttered scene. Also, our representations lend themselves to applications other than classification (e.g. manipulation).

2   Hierarchical Model

Our object model consists of a set of generic *features* organized in a hierarchy. Features that form the bottom level of the hierarchy, referred to as *primitive features*, are bound to visual observations. The rest of the features are *meta-features* which embody spatial configurations of more elementary features, either meta or primitive. Thus, a meta-feature incarnates the relative configuration of two features from a lower level of the hierarchy.

A feature can intuitively be associated to a "part" of an object, i.e. a generic component instantiated once or several times during a "mental reconstruction" of the object. At the bottom of the hierarchy, primitive features correspond to local parts that each may have many *instances* in the object. Climbing up the hierarchy, meta-features correspond to increasingly complex parts defined in terms of constellations of lower parts. Eventually, parts become complex enough to satisfactorily represent the whole object.

Formally, the hierarchy is implemented in a Pairwise Markov Random Field. Features correspond to hidden nodes of the network. When a model is associated to a particular scene (during construction or instantiation), the pose of feature i in that scene will be represented by the probability density function of the random variable x_i associated to feature i, effectively linking feature i to its instances. Random variables are thus defined over the pose space $SE(3) = \mathbb{R}^3 \times SO(3)$.

The structure of the hierarchy is reflected by the edge pattern of the network; each meta-feature is thus linked to its two child features. As noted above, a meta-feature encodes the relationship between its two children. However, the graph records this information in a slightly different but equivalent way: instead of recording the relationship between the two child features, the graph records the two relationships between the meta-feature and each of its children. The relationship between a meta-feature i and one of its children j is parametrized by a *compatibility potential function* $\psi_{ij}(x_i, x_j)$ associated to the edge e_{ij}. A compatibility potential specifies, for any given pair of poses of the features it links, the probability of finding that particular configuration for these two features. We only consider rigid-body relationships. Moreover, relationships are *relative* spatial configurations. Compatibility potentials can thus be represented by a probability density over the feature–to–feature transformation space $SE(3)$.

Finally, each primitive feature is linked to an observed variable y_i. Observed variables are tagged with an appearance descriptor called a *codebook vector*. The set of all codebook vectors forms a *codebook* that binds the object model to feature observations. The statistical dependency between a hidden variable x_i and its observed variable y_i is parametrized by an *observation potential* $\phi_i(x_i)$, also referred to as *evidence* for x_i, which corresponds to the spatial distribution of the observations. We generally cannot observe meta-features; their observation potential is thus uniform.

3 Inference

Model instantiation is the process of detecting instances of an object model in a scene. It provides pose densities for all features of the model, indicating where the learned object is likely to be present. Instantiating a model in a scene amounts to inferring posterior marginal densities for all features of the hierarchy.

The first step of inference is to define priors (observation potentials, evidence) for all features (hidden nodes) of the model. For primitive features, evidence is estimated from feature observations. Observations are classified according to the observation codebook; for each primitive feature i, its observation potential $\phi_i(x_i)$ is estimated from observations that are (softly) associated to the ith codebook vector. For meta-features, evidence is uniform.

Once priors have been defined, instantiation can be achieved by any applicable inference algorithms. We currently use a Belief Propagation algorithm of which we give a complete, top-down view below.

3.1   Belief Propagation

Belief Propagation (BP) [7,10,13] is based on incremental updates of marginal probability estimates, referred to as *beliefs*. The belief at feature i is denoted by

$$b(x_i) \approx \mathbf{P}(x_i|y) = \int ... \int \mathbf{P}(x_1, ..., x_N|y)\, dx_1...dx_{i-1}dx_{i+1}...dx_N$$

where y stands for the set of observations. During the execution of the algorithm, *messages* are exchanged between neighboring features (hidden nodes). A message that feature i sends to feature j is denoted by $m_{ij}(x_j)$, and contains feature i's belief about the state of feature j. In other words, $m_{ij}(x_j)$ is a real positive function proportional to feature i's belief about the plausibility of finding feature j in pose x_j. Messages are exchanged until all beliefs converge, i.e. until all messages that a node receives predict a similar state.

At any time during the execution of the algorithm, the current pose belief (or marginal probability estimate) for feature i is the normalized product of the local evidence and all incoming messages, as

$$b_i(x_i) = \frac{1}{Z}\phi_i(x_i) \prod_{j\in\text{neighbors}(i)} m_{ji}(x_i), \tag{1}$$

where Z is a normalizing constant. To prepare a message for feature j, feature i starts by computing a "local pose belief estimate", as the product of the local evidence and all incoming messages *but* the one that comes from j. This product is then multiplied with the compatibility potential of i and j, and marginalized over x_i. The complete message expression is

$$m_{ij}(x_j) = \int \psi_{ij}(x_i, x_j)\phi_i(x_i) \prod_{k\in\text{neighbors}(i)\backslash j} m_{ki}(x_i)dx_i. \tag{2}$$

As we see, the computation of a message doesn't directly involve the complete local belief (1). In general, the explicit belief for each node is computed only once, after all desirable messages have been exchanged.

When BP is finished, collected evidence has been propagated from primitive features to the top of the hierarchy, permitting inference of the top feature marginal pose density. Furthermore, regardless of the propagation scheme (message update order), the iterative aspect of the message passing algorithm ensures that global belief about the object pose – concentrated at the top nodes – has at some point been propagated back down the hierarchy, reinforcing globally consistent evidence and permitting the inference of occluded features. While there is no theoretical proof of BP convergence for loopy graphs, empirical success has been demonstrated in many situations.

3.2 Nonparametric Representation

We opted for a nonparametric approach to probability density representation for all entities of the model, i.e. random variable and functions of random variables, including potentials, messages, and evidence. A density is simply represented by a set of (possibly weighted) particles; the local density of these particles in a given region is proportional to the actual probabilistic density in that region. The number of particles supporting a density is fixed, and will be denoted by n. Whenever a density has to be evaluated, traditional kernel density estimation methods can be used. Compared to usual parametric approaches that involve a limited number of parametrized kernels, a nonparametric approach eliminates problems like fitting of mixtures or the choice of a number of components. Also, no assumption concerning the shape of the density has to be made.

Figure 1 shows an example of a hierarchy for a traffic sign. Feature 2 is a primitive feature that corresponds to a local black-white edge segment – the white looks greenish on the picture. The blue patch pattern in the $\phi_2(x_2)$ box is the non-parametric representation for the evidence distribution for feature 2. The blue patch pattern in the x_2 box is the non-parametric representation for the posterior density of x_2, i.e. the poses in which part "feature 2" is likely to be found. Feature 4 is the combination of primitive features 1 and 2. The red patch in the x_4 box shows its inferred pose in the scene. The $\psi_{4,2}(x_4, x_2)$ box shows the encoding of the relationship between features 4 and 2; for a fixed pose for feature 4 (in red, bottom right of the box), it shows the likely poses for feature 2 (in blue). The sign itself corresponds to feature 6, denoted by its random variable x_6. It is the composition of two features, one representing the central "opening bridge" pattern *and* the corners of the inner triangle (feature 4), the other representing the central pattern *and* the outer edges (feature 5).

3.3 Nonparametric Belief Propagation

For inference, we use a variant of BP, Nonparametric Belief Propagation (NBP), an algorithm for BP message update in the particular case of continuous, non-Gaussian potentials [12]. The underlying method is an extension of particle filtering; the representational approach is thus nonparametric and fits our model very well.

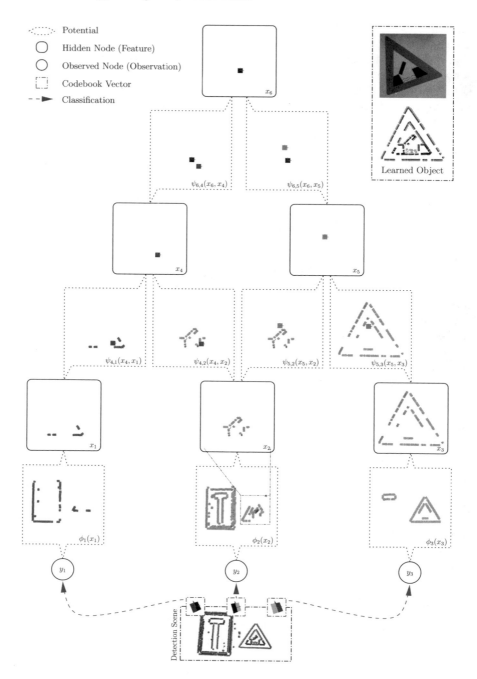

Fig. 1. Example of a hierarchical model of a traffic sign

NBP is easier to explain if we decompose the analytical message expression (2) into two steps:

1. Computation of the local belief estimate

$$\beta_{ts}(x_t) = \phi_t(x_t) \prod_{i \in N(t) \backslash s} m_{it}(x_t), \qquad (3)$$

2. Combination of β_{ts} with the compatibility function ψ_{ts}, and marginalisation over x_t

$$m_{ts}(x_s) = \int \psi_{ts}(x_t, x_s) \beta_{ts}(x_t) dx_t. \qquad (4)$$

NBP forms a message by first sampling from the product (3) to collect a non-parametric representation for $\beta_{ts}(x_t)$, it then samples from the integral (4) to collect a non-parametric representation for $m_{ts}(x_s)$. These two operations are executed alternately: transform local estimate to form a message, merge messages to form a local estimate, etc...

Sampling from the message product (3) is conceptually straightforward. Using Gaussian kernel density estimation, each factor (messages and evidence) can be represented by a weighted sum of n Gaussians. The product of a series of Gaussians is also a Gaussian, and the parameters (mean, variance, weight) of the product Gaussian can easily be computed from the parameters of the factor Gaussians. Hence, letting $d = (N(t) - 1) + 1$ denote the number of factors in the product (3), $\beta_{ts}(x_t)$ can be expressed as a weighted sum of n^d Gaussians [12]. A nonparametric representation for $\beta_{ts}(x_t)$ can thus be constructed by sampling from a mixture of n^d Gaussians, which amounts to repetitively selecting one Gaussian at random and taking a random sample from it. The computational cost of this exhaustive approach is $O(n^d)$. Clearly, exhaustive product implementations will suffer from overly long computation times.

The second phase of the NBP message construction computes an approximation for the integral (4) by stochastic integration. Stochastic integration takes a series of samples $\hat{x}_t^{(i)}$ from $\beta_{ts}(x_t)$, and propagates them to feature s by sampling from $\psi_{ts}(\hat{x}_t^{(i)}, x_s)$ for each $\hat{x}_t^{(i)}$. It would normally also be necessary to take into account the marginal influence of $\psi_{ts}(x_t, x_s)$ on x_t. In our case however, potentials only depend on the difference between their arguments; the marginal influence is a constant and can be ignored.

3.4   Importance Sampling

The computational bottleneck of NBP clearly lies in message products. Ihler et al. explored multiple improvements over the exhaustive product [6], one of which is to sample from the product using Importance Sampling (IS). IS is a technique for sampling from an unknown distribution $p(x)$ by sampling a series of examples $\hat{x}^{(\ell)}$ from a known distribution $q(x)$ ideally similar to p. IS accounts for the difference between the target distribution p and the proposal distribution q by assigning to each sample a weight defined as

$$w^{(\ell)} = \frac{p(\hat{x}^{(\ell)})}{q(\hat{x}^{(\ell)})}.$$

To produce a sample of size n, one usually takes rn weighted examples from q, where $r > 1$, and eventually resamples them to a size of n. The closer q is to p, the better $\{\hat{x}^{(\ell)}\}$ will approximate p.

Sampling from a message product (3) with IS works by selecting one of the messages $m_{ut}(x_t)$ (or the evidence) as proposal distribution, the rest of the factors providing importance weights:

$$w^{(\ell)} = \frac{\phi_t(\hat{x}_t^{(\ell)}) \prod_{i \in N(t) \backslash s} m_{it}(\hat{x}_t^{(\ell)})}{m_{ut}(\hat{x}_t^{(\ell)})} = \phi_t(\hat{x}_t^{(\ell)}) \prod_{i \in N(t) \backslash \{s,u\}} m_{it}(\hat{x}_t^{(\ell)}).$$

IS produces n samples from a product of d factors in $O(rdn^2)$ time. From here on, we will consider that the number of neighbors a node may have is bounded and typically low, and ignore it in complexity statements. IS thus produces n samples from a product of d factors in $O(rn^2)$ time.

4   Efficient Importance Sampling of Message Products

The success of NBP inference highly depends on a sufficient density resolution, i.e. having enough particles to support the different modes of potentials, local estimates, and messages. Moving to more complex applications will generally require an increase of n, which has a hard impact on computational time and memory needs. This section presents a variant of the IS-based NBP algorithm that yields a significant improvement of the inference power without any memory impact. Its computational behavior is close to original IS-based NBP, with some interesting benefits.

4.1   Representational Constraints

As explained above, *A message that feature i sends to feature j – denoted by* $m_{ij}(x_j)$ *– contains feature i's belief about the state of feature j*. Feature i will often possess a rather inaccurate local estimate, e.g. at the beginning of propagation when each bottom feature receives observations from the whole scene surrounding an object of interest. Additionally, even if a local estimate was exact, transforming it with ψ_{ij} will generate a large number of possible states for feature j, only a fraction of which will eventually become confirmed by other messages incoming to j – the job of message products precisely is to extract sections that overlap between incoming messages. Generating a message from local estimates can be pictured as an exploration process, while merging messages together would be a confirmation/concentration process. From these observations, it intuitively follows that one may achieve better performance by increasing the resolution of messages only, leaving potentials and local estimates at their initial resolution.

4.2   Implicit Messages

Let us now turn to the propagation equation (2), which we *analytically* decomposed into a multiplication (3) and an integration (4). We explained that NBP implements BP by *physically* performing the same decomposition, i.e. computing explicit nonparametric representations for messages and local estimates alternately. In this section, we propose a somewhat different implementation, in which explicit representations are only computed for local estimates.

Let us assume we are in the process of constructing a nonparametric representation for $\beta_{ts}(x_t)$, i.e. the local estimate of feature t that includes all incoming information but that from s. In typical IS-based NBP, we first choose one incoming message $m_{ut}(x_t)$ at random ($u \neq s$) as IS proposal density; then, we repetitively take a sample $\hat{x}_t^{(\ell)}$ from $m_{ut}(x_t)$ and compute its importance weight

$$w^{(\ell)} = \phi_t(\hat{x}_t^{(\ell)}) \prod_{i \in N(t) \backslash \{s,u\}} m_{it}(\hat{x}_t^{(\ell)}). \tag{5}$$

One can notice though that neither of these two operations do actually need an explicit expression for incoming messages. Producing $\hat{x}_t^{(\ell)}$ from $\beta_{ut}(x_t)$ and $\psi_{ut}(x_u, x_t)$ is straightforward. In turn, Expression (5) can be rewritten

$$w^{(\ell)} = \phi_t(\hat{x}_t^{(\ell)}) \prod_{i \in N(t) \backslash \{s,u\}} \int \psi_{it}(x_i, \hat{x}_t^{(\ell)}) \beta_{it}(x_i) dx_i. \tag{6}$$

Evaluating each integral is achieved by sampling p times an example $\hat{x}_i^{(k)}$ from either $\psi_{it}(x_i, \hat{x}_t^{(\ell)})$ or $\beta_{it}(x_i)$, evaluating $\beta_{it}(\hat{x}_i^{(k)})$ or $\psi_{it}(\hat{x}_i^{(k)}, \hat{x}_t^{(\ell)})$ respectively, and taking the average over k.

The computational complexity of importance weight computation with explicit messages (5) is $O(n)$, because of linear iteration through all messages and evidence which are of size n. The computational complexity with implicit messages (6) is $O(pn)$, because of p linear iterations through potentials or the local estimates. However, implicit messages effectively achieve the same resolution as explicit messages would *if* these explicit messages were supported by pn particles, *while keeping memory needs at $O(n)$*. Importance weight computation with implicit or explicit messages are thus expected to display processing times of the same order, while the implicit method will categorically require less memory.

4.3   Two-Level Importance Sampling

One known weakness of IS-based NBP is that it cannot intrinsically concentrate its attention on the modes of a product, which is an issue since individual messages often present many irrelevant modes [6]. We overcome this problem with a two-level IS: we first compute an intermediate representation for the product with the procedure explained above, we then use this very representation as the proposal distribution for a second IS that will be geared towards relevant modes. The intermediate representation is obtained with sparse implicit messages ($p \ll n$) but many importance samples ($r \gg 1$), while the second IS uses rich implicit messages ($p \approx n$) but a low value for r. Denoting by $\beta_{ts}^*(x_t)$ the intermediate product representation, importance weights for the second IS are computed as

$$w^{(\ell)} = \frac{\phi_t(\hat{x}_t^{(\ell)}) \prod_{i \in N(t) \backslash s} m_{it}(\hat{x}_t^{(\ell)})}{\beta_{ts}^*(\hat{x}_t^{(\ell)})}.$$

In the equation above, messages are implicit.

The two-level IS described above and the high-resolution messages have been crucial elements of the successful application to real-world object presented at Section 5.2.

5   Evaluation

5.1   Pose Estimation

The feature at the top of a hierarchical object model represents the whole object. When instantiating the model in a scene in which exactly one instance of the object is present, the top feature density should present one major mode, which can be used to estimate the object pose. Let us consider a model for a given object, and a pair of scenes where the object appears. In the first scene, the object is in a reference pose. In the second scene, the pose of the object is unknown. The application of our method to estimate the pose of the object in the second scene goes as follows:

1. Instantiate the object model in the reference scene, and compute a *reference object pose* π_1 as the mean of the top feature density major mode.
   We emphasize that a hierarchy comes from *unsupervised* recursive combinations of features [2]. Even though the *object* is in a reference pose, π_1 is not expected to be located at $(0,0,0)$ or aligned with $(\mathbf{x},\mathbf{y},\mathbf{z})$, which makes this first step necessary.
2. Instantiate the object model in the unknown scene and compute pose π_2 from the major mode of the top feature density.
3. Let t be the transformation between π_1 and π_2. This transformation corresponds to the rigid body motion between the pose of the object in the first scene and its pose in the second scene. Since the first scene is a reference pose, t is the *pose* of the object in the second scene.

A prominent aspect of this procedure is its ability to recover an object pose without explicit point-to-point correspondences. The estimated pose emerges from a negotiation involving all available data.

5.2   Experiments

In this section, we demonstrate the applicability of our model with a series of pose estimation experiments in various cluttered scenes. We chose to learn models for the three objects presented at Figure 2(a). We then tried to estimate their poses in the scenes of Figure 2(b).

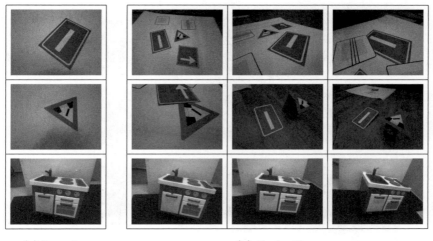

<table>
(a) Learning   (b) Evaluation
</table>

Fig. 2. Input imagery (only the left image in each stereo pair is presented). Effective resolution is 1280×960 pixels.

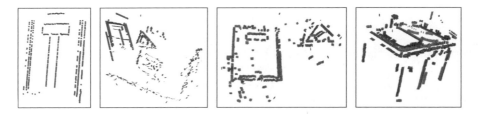

Fig. 3. Examples of ECV representations, extracted from scenes of Figure 2

Observations are provided by an early-cognitive-vision (ECV) system [8], which extracts 3D primitives from stereo views of a scene. The quality of such ECV representations varies as a function of local visual signal quality. Figure 3 illustrates the ECV primitives for certain scenes of Figure 2.

Models for the three objects of Figure 2(a) were learned following the procedure mentioned above [2]. These models were learned from a clean view of each object (the reference scene), for example from the ECV representation in the first image of Figure 3. Each model has also been instantiated in its reference scene to compute its reference pose π_1.

The three models were all instantiated in the test scenes of Figure 2(b), using observations like these of Figure 3 as evidence. Looking closer at the instantiation of one model in one scene, there are two cases to consider. First, the model had no instance in the scene. The top-feature density was then relatively uniform, and the experiment did not need to go any further. In the second case, an instance was present. It was then always verified that the top feature did present a principal mode π_2. We could thus compute the transformation t between π_1 and π_2, which

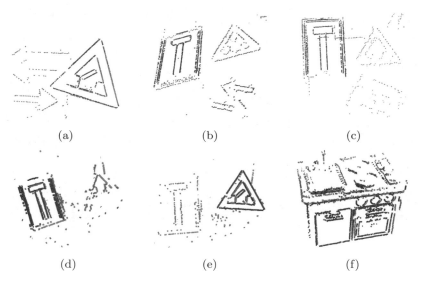

(a)   (b)   (c)

(d)   (e)   (f)

Fig. 4. Illustration of the pose estimation accuracy. Each picture shows in green a scene that contains one object of interest and in red the pose of that object inferred by our system.

corresponds to the *estimated* rigid body motion between the pose of the object in the reference scene, and to its pose in the noisy scene.

We can evaluate the success of the experiment by transforming the reference scene with t, and superimposing it onto the test scene; if the experiment is successful, the object of interest should overlap with its instance. Such evaluations are presented at Figure 4. All the experiments that we ran ended with successful pose recovery. For traffic signs, the worst estimate (Figure 4(d)) corresponds to the dead-end signal pose estimation in the sixth scene of Figure 2(b) (second row, third column). This is however one of the most difficult scenes: it has a brown background, thus changing the outside color of ECV primitives on the traffic sign contours. This induces wrong associations of observations to primitive features, and makes for harder inference. Estimation is still quite accurate given the difficulty of the scene. Other typical estimates are presented at Figure 4. In particular, 4(a) shows a good result despite occlusion.

The accuracy of probabilistic pose estimation highly depends on the resolution of the representation. When an experiment lacks accuracy, retrying with more particles usually produces better results. Therefore, a meaningful quantitative evaluation must take into account the number of particles per density. Figure 5 shows the pose estimation error as a function of the number of particles per density. Because of the probabilistic nature of inference, runs with different software random seeds produce different results. Therefore, we run each experiment several times and study the mean error, plotted in red in the figure. The mean error decreases quickly when going from 40 to 100 particles, and stabilizes for

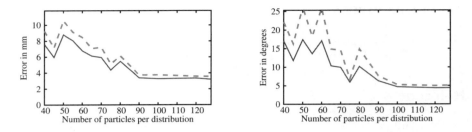

Fig. 5. Pose estimation accuracy as a function of the number of particles per density, for an instantiation of the opening-bridge traffic sign within the first scene of Figure 2(b). Left and right plots correspond to location and orientation error respectively. The solid lines (red) indicate the mean absolute error. The dashed lines (green) indicate the variance across runs. Location error can be compared to the traffic sign edge, which is 190mm long. See the text for details.

higher resolutions. We also plotted one standard deviation above the mean error, in dashed green. The error variance also decreases as the number of particles increases.

6   Discussion

6.1   Related Work

Compared to recent work in the field [1,3,4], the most distinguishing aspects of our approach are its explicit 3D support and the unbiased contributions of all low-level features. We learn from observations defined in 3D, and infer a full 3D pose. The use of a sophisticated inference algorithm permits the uniform integration of all available evidence, avoiding an explicit combinatorial search.

6.2   Conclusion

We presented an object representation framework that encodes probabilistic relations between 3D features. We discussed an Importance-Sampling–NBP inference process which, together with the learning scheme of our previous work [2], allow us to learn unsupervised part representations for real objects and to instantiate them in cluttered scenes. We are thus able to achieve pose recovery without prior object models, and without explicit point correspondences.

Our method can in principle incorporate features from more perceptual modalities than vision. Our objective is to observe haptic and kinematic features that correlate with successful grasps, and integrate them into the feature hierarchy. Then, given a visual scene, grasp parameters can be suggested by probabilistic inference within the feature hierarchy.

Acknowledgment

This work was supported by the Belgian National Fund for Scientific Research (FNRS) and the EU Cognitive Systems project PACO-PLUS (IST-FP6-IP-027657).

References

1. Bouchard, G., Triggs, B.: Hierarchical part-based visual object categorization. Computer Vision and Pattern Recognition 1, 710–715 (2005)
2. Detry, R., Piater, J.H.: Hierarchical integration of local 3D features for probabilistic pose recovery. In: Robot Manipulation: Sensing and Adapting to the Real World (Workshop at Robotics, Science and Systems) (2007)
3. Epshtein, B., Ullman, S.: Feature hierarchies for object classification. In: IEEE International Conference on Computer Vision (2005)
4. Fidler, S., Leonardis, A.: Towards scalable representations of object categories: Learning a hierarchy of parts. In: CVPR 2007 (2007)
5. Fukushima, K.: Neocognitron: A self-organizing neural network model for a mechanism of pattern recognition unaffected by shift in position. Biological Cybernetics 36(4), 193–202 (1980)
6. Ihler, A.T., Sudderth, E.B., Freeman, W.T., Willsky, A.S.: Efficient multiscale sampling from products of Gaussian mixtures. In: Neural Information Processing Systems (2003)
7. Jordan, M.I., Weiss, Y.: Graphical models: Probabilistic inference. In: Arbib, M. (ed.) The Handbook of Brain Theory and Neural Networks, 2nd edn. MIT Press, Cambridge (2002)
8. Krüger, N., Wörgötter, F.: Multi-modal primitives as functional models of hypercolumns and their use for contextual integration. In: De Gregorio, M., Di Maio, V., Frucci, M., Musio, C. (eds.) BVAI 2005. LNCS, vol. 3704, pp. 157–166. Springer, Heidelberg (2005)
9. Marr, D., Nishihara, H.K.: Representation and recognition of the spatial organization of three dimensional shapes. Proceedings of the Royal Society of London B 200, 269–294 (1978)
10. Pearl, J.: Probabilistic Reasoning in Intelligent Systems: Networks of Plausible Inference. Morgan Kaufmann, San Francisco (1988)
11. Riesenhuber, M., Poggio, T.: Hierarchical models of object recognition in cortex. Nat. Neurosci. (1999)
12. Sudderth, E.B., Ihler, A.T., Freeman, W.T., Willsky, A.S.: Nonparametric belief propagation. In: CVPR 2001, p. 605 (2003)
13. Yedidia, J.S., Freeman, W.T., Weiss, Y.: Understanding belief propagation and its generalizations. Technical report, Mitsubishi Electric Research Laboratories (2002)

Semantic Reasoning for Scene Interpretation

Lars B.W. Jensen[1], Emre Baseski[1], Sinan Kalkan[2], Nicolas Pugeault[3], Florentin Wörgötter[2], and Norbert Krüger[1]

[1] University of Southern Denmark
Odense, Denmark
{lbwj,emre,norbert}@mmmi.sdu.dk
[2] University of Göttingen
Göttingen, Germany
{sinan,worgott}@bccn-goettingen.de
[3] University of Edinburgh
Edinburgh, United Kingdom
npugeaul@ed.ac.uk

Abstract. In this paper, we propose a hierarchical architecture for representing scenes, covering 2D and 3D aspects of visual scenes as well as the semantic relations between the different aspects. We argue that labeled graphs are a suitable representational framework for this representation and demonstrate its potential by two applications. As a first application, we localize lane structures by the semantic descriptors and their relations in a Bayesian framework. As the second application, which is in the context of vision based grasping, we show how the semantic relations can be associated to actions that allow for grasping without using any object knowledge.

Keywords: cognitive vision, semantic reasoning, bayesian classification.

1 Introduction

In this work, we represent scenes with a hierarchy of visual information. The input consists of stereo images (or sequences of them) that become processed at different levels. Information of increasing semantic richness becomes processed at the different levels, covering multiple aspects of a scene such as 2D and 3D information as well as geometric and appearance based information. Furthermore, the spatial extent of the processed entities increases in the higher levels of the hierarchy.

We make use of rich local symbolic descriptors, describing edge-like structures and homogeneous structures, as well as groups (contours and areas) formed by them. Furthermore, rich semantic relations between these descriptors and the groups are defined. The descriptors describe local information in terms of multiple visual modalities (2D and 3D position and orientation, colour as well as contrast transition). Moreover, there is a set of semantic relations defined between them such as the Euclidean distance in 2D and 3D as well as parallelism, co-planarity and co-colority (i.e., sharing similar colour structure).

B. Caputo and M. Vincze (Eds.): ICVW 2008, LNCS 5329, pp. 121–134, 2008.

Scenes become represented as a set of labeled graphs, whose nodes are labeled by properties of local descriptors, groups and areas thereof and edges between the nodes represent the semantic relations between the nodes in the graphs. Idealized graphs can be defined or learned from scene structures such as road lanes and can be efficiently matched with the extracted scene graphs by making use of the rich semantics.

From a cognitive point of view, it is important to have a representation that allows for an efficient storage of information as well as for reasoning processes on visual scenes. From a storage point of view, it is not convenient to memorize information on a very low and local level since it would require a large amount of memory. Also it would be much more difficult for learning processes to make use of relevant semantics. As a consequence, the very condensed graph representation is much more suitable for memorizing objects.

We present two applications of our hierarchical framework: As a first application, we show how a street structure can be characterized by both its appearance and relations between its sub-components. Here, the matching process is governed by Bayesian reasoning based on local descriptors and semantic relations between them, which are controlled by prior probabilities. Moreover, this Bayesian reasoning process makes explicit the relative importance of the different cues and relations opening the way for the learning of sparse graph structures. In terms of semantic reasoning, we can show that, by means of the semantic relations, it is possible to mediate between textual descriptions of scene structures (e.g., the lanes) and visual detection as examplified. Such graphs can be idealized (or, generalized) either through learning or can be provided as world knowledge, and be used for matching (see section 4.1).

The second application is based on [1], and illustrates how the approach presented herein embeds in a robotic scenario. In this scenario, groups of visual features fulfilling certain semantic relations can be associated to grasping actions, allowing for the grasping of objects without using any model knowledge.

The use of hierarchical representations, mostly graphs, is commonplace for scene representation. For example, *scene graphs* and *spatial relationship graphs* are heavily used in Computer Graphics for representing 3D world and scenes [2]; such graphs are designed mostly for rendering purposes, and they are not sufficient for covering the 2D properties of scenes. *Relative Neighborhood Graphs*, introduced by [3], are used in Computer Vision studies for representation of structured entities [4]. A similar graphical structure called *Region Adjacency Graph* is used for region-based representation of objects or scenes [5,6]. There exist a variety of similar graphical representations and we refer the interested reader to [7].

Our contribution in this paper is the introduction of a hierarchical vision system that allows for semantic reasoning based on rich descriptors and their relations. This vision system covers not only the appearance aspects but also the geometrical properties of the scene, which allows for doing reasoning in both 2D and 3D world. In particular, it allows for the step-wise translation of a textual description of an object to a visual representation that can be used for localizing a certain structure in a visual scene.

The paper is structured as follows: In section 2, the visual scene representation is introduced. In section 3, we describe the embedding of the visual representation in graphs. We then describe the two applications in section 4. We introduce the algorithm for the detection of a lane structure in section 4. Another application in the context of vision based grasping is described in section 4.2. In section 5, we discuss the potential of this approach in terms of a cognitive system architecture.

2   Hierarchical Architecture

We represent scenes with a three-level architecture of visual entities (see figure 1) of increasing richness and semantic. In the following subsections, we introduce the different levels of this hierarchical representation in order of increasing complexity, starting from the lowest level.

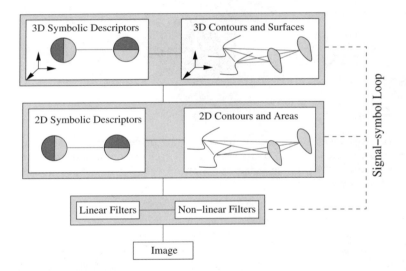

Fig. 1. An overview of the hierarchical architecture introduced in this paper. The visual entities denote the nodes of the graphical representation, and the red edges, which correspond to perceptual grouping and correspondence relations, are the links between the nodes. Higher levels in the hierarchy correspond to more symbolic, more spacious and more descriptive visual entities. See the text for more details and figure 6 for examples of the different levels of the hierarchical architecture.

2.1   Linear and Non-linear Filtering

At the first level, we apply a combination of linear and non-linear filtering operations to extract pixel-wise signal information in terms of local magnitude, orientation, phase [8] as well as optical flow [9] — for details see [10,11].

2.2   Symbolic Representation in 2D

The transition to a local symbolic description is done at the second level (the "Symbolic Representation in 2D" layer in figure 6) where local image patches are described by the so-called *multi-modal primitives* [12]. The primitives provide a condensed semantic description of the local (spatial-temporal) signal in terms of image orientation, phase, colour and optic flow. The difference to the first level is that the information is sparsified, highly condensed and associated to discrete

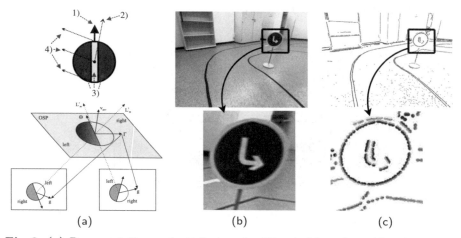

(a)   (b)   (c)

Fig. 2. (a) Representation and attributes of a 2D primitive where (1) orientation of the primitive, (2) the phase, (3) the color and (4) the optic flow and reconstruction of a 3D primitive. **(b)** A sample scene and a closer view for the region of interest. **(c)** Extracted 2D primitives for the example scene in **(b)**.

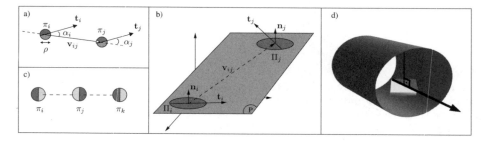

Fig. 3. Illustration of the perceptual relations between primitives. **(a)** Collinearity of two 2D primitives. **(b)** Co–planarity of two 3D primitives Π_i and Π_j. **(c)** Co–colority of three 2D primitives π_i, π_j and π_k. In this example, π_i and π_j are cocolor, so are π_i and π_k; however, π_j and π_k are not cocolor. **(d)** Normal distance between Π_i and Π_j is 0 if Π_j is outside the cylindrical volume surrounding Π_i and defined otherwise as the distance between Π_j and the line created from the location of Π_i which goes in the direction of Π_is orientation vector.

positions with sub-pixel accuracy. Figure 2 shows extracted 2D primitives (denoted as π) for an example scene.

At this level, the information is sparsely coded such that interaction processes between visual events can be modeled more efficiently than at the pixel level (for a detailed description of these interaction processes see, e.g., [13]). Already at this level, semantic relations between local 2D primitives can be defined. Besides the 2D distance, primitives allow collinearity and co-colourity relations to be defined between them: Two primitives are collinear if they are part of the same line (figure 3(a) and 4(f)). Two primitives, on the other hand, are co-colour if the colours of their *sides* that face each other are similar (figures 3(c) and 4(e)). See [14] for more information about the definition of these relations.

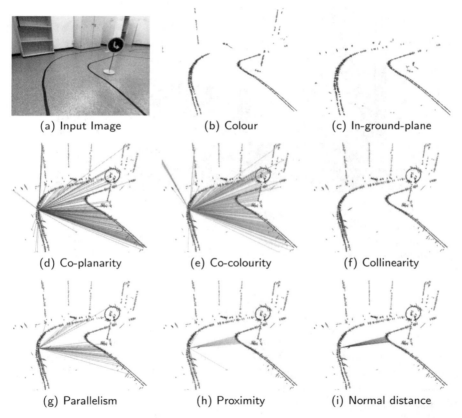

(a) Input Image   (b) Colour   (c) In-ground-plane

(d) Co-planarity   (e) Co-colourity   (f) Collinearity

(g) Parallelism   (h) Proximity   (i) Normal distance

Fig. 4. A set of 2D and 3D relations for the visual entities extracted from an example scene whose left view is provided in (a). (b) Primitives which are black. (c) 3D primitives which satisfy the "ground-plane" relation. (d-g) Connects the 3D primitives that are respectively co-planar, co-colour, collinear and parallel to a selected 3D primitive. (h) Connects the 3D primitives that are of a given 3D distance to a selected 3D primitive. (i) Connects the 3D primitives whose normal distance to a selected 3D primitive equals a given value.

The 2D descriptors naturally organize themselves along contours and the semantic description is highly correlated along such a contour (e.g., 2D orientation varies smoothly and in general colour, phase and optic flow are similar for the primitives on the contour). Hence, it is natural to condense the information of the primitives organized along a contour in the form of a more abstract parameterization in terms of unified appearance based descriptors as well as a NURBS (Non-Uniform Rational B-Splines [15]) representation of the geometry of the contours (see figure 5). By this, we reduce the number of bits used to represent a scene further as well as the number of second order relations of visual events. The latter point is in particular relevant, when we want to code objects with these relations.

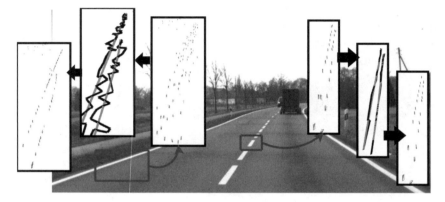

Fig. 5. Position and orientation correction of 3D primitives by using NURBS. After fitting NURBS (represented as green lines) to groups of primitives (represented as black lines), position and orientation of each primitive is recalculated. The procedure is shown on a good reconstruction (middle road marker) as well as a bad one (left lane marker).

2.3   Symbolic Representation in 3D

Using the corresponding 2D primitives in the left and right image, 3D primitives can be reconstructed (denoted by $\boldsymbol{\Pi}_j$). At the third level, the reconstructed 3D primitives inherit the appearance based properties of the 2D primitives (phase and colour) and extend the 2D position and 2D orientation to 3D (see figure 2). Moreover, the semantic relations between 2D primitives can be extended to the 3D primitives and also further enriched by particular 3D relations such as co-planarity or 3D properties such as in-ground-plane (see figures 3 and 4). Co-planarity refers to the *being-on-the-same-plane* relation between two 3D primitives or 3D contours (figures 3(b) and 4(d)). See [14] for more information about the definition of co-planarity. In-ground-plane relation, on the other hand, corresponds to all 3D entities that are in the ground plane (figure 4(c)). The 2D contour representation becomes also extended to 3D contours by connecting 3D primitives that are linked together. NURBS are fitted to the 3D contours as in 2D to obtain a global mathematical description of the 3D contours. In addition, the NURBS parametrization can be used to increase the precision of the local feature extraction process (see figure 5).

Note that this process is not a pure bottom-up process, as it involves corrective feedback mechanisms at various levels. These are described in more detail in, e.g., [13,16].

3 Semantic Graphs

The hierarchy of representations discussed above provides us with a number of 2D and 3D local entities that are linked to more global entities. These entities are semantically rich as such, and in addition there exist semantic relations between them. Because of this linkage, we suggest that labeled graphs are the suitable representational framework for representing scenes. In these graphs, the nodes represent different visual entities such as primitives, contours and areas with their first order properties while the links represent the semantic relations. Note that actually we have a set of labeled graphs, which are linked to each other and with this linkage, they cover the 2D and 3D aspects of a scene (see figure 6) since each relation naturally defines a sub-graph, covering a structure in a scene.

In processing of information across the different levels, the semantic richness of information increases from level to level. However, it is important to point out that with this increase of semantical richness, also the likelihood of errors in the processing increases due to loss of valuable information or introduction of noise through thresholding. In addition, the uncertainty of visual information, in particular in the 3D domain, might also make any reasoning uncertain. Hence, we intend to be able to use the extracted information *on all levels* according to the current task and uncertainties of information at the different levels. In addition, spatial–temporal processes are defined that increase the stability and the certainty of information by spatial-temporal predictions [13]. The proposed hierarchy allows for processes that transfer information from the symbolic level to the signal level to recover weak information in so-called signal-symbol loops (see [16]). Such loops are essentially feedback mechanisms that carry the results of symbolic processing to the signal level.

4 Applications

In this section, we give two applications of the semantic reasoning process. First, we show how a lane structure can be described by the semantic descriptors and their relations in a Bayesian framework (section 4.1). Then we describe another application in a robotic context (section 4.2).

4.1 Lane Finding Using Bayesian Reasoning

A lane in our lab environment (see figure 4(a)) can be characterized by the colour and the width of the lane marker, which is known also to be in the ground plane,

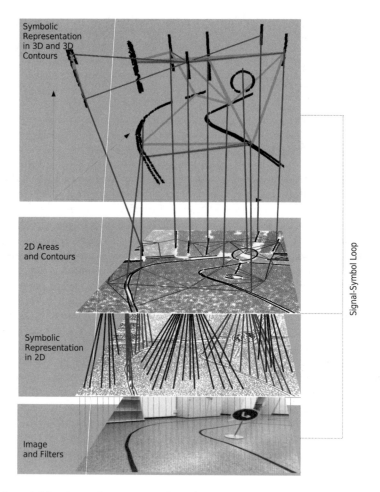

Fig. 6. A multi-level graph structure. For clarity, only a subset of the links is drawn, and the links corresponding to different relations such as parallelism and co-colority between 2D or 3D entities are skipped. "Image and Filters" (IF) layer is the input image which contains pixels as the nodes of the graph. "Symbolic Representation in 2D" (SR-2D) layer contains the 2D primitives. The links between the IF layer and the SR-2D layer correspond to "part-of" relations between pixels and primitives. "2D Contours and Areas" (CA-2D) layer contains image areas (each area is drawn in a different color) and 2D contours (in black). The neighborhood relations between two areas and between an area and a contour are drawn respectively in blue and red. The links between the SR-2D layer and the CA-2D layer correspond to "part-of" relations between primitives, and areas and contours. The "Symbolic Representation in 3D and 3D Contours" (SRC-3D) layer includes 3D contours in black (the 3D surfaces are skipped for clarity), and the links in red and light green between the 3D contours respectively denote coplanarity and cocolority relations between the contours. The links between the CA-2D layer and the SRC-3D layer are "projection" relations between the 2D and 3D contours.

as well as by its distance to the other lane marker. As a textual description of the lane one could state:

A lane consists of two lane markers with distance d_{far} which are both in the ground plane. A lane marker has a width d_{near} and has the colour 'black'.

An idealized representation of this textual description in a graph is shown in figure 7. The representation introduced in the last two sections allows for directly applying the terms used in the textual description. Colour and 'being in ground plane' are first order attributes of primitives and groups while the term 'distance' corresponds to the relation 'normal distance' (figure 3). Hence, the textual description can be easily translated in our visual representations. However, there are two problems we have to face: First, a lane is not described by one property, or relation, but by a number of properties. Therefore, these different cues need to be combined. Second, scene interpretation processes have to face uncertainties in the feature extraction process. Reasons for the uncertainties are, for example, noise in the recording process, limited resolution as well as the correspondence problem in the stereo reconstruction.

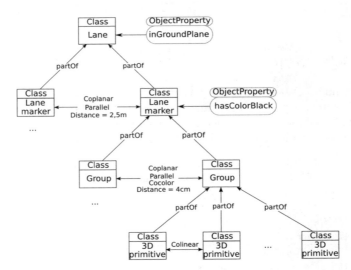

Fig. 7. A graph showing an idealized representation of the lane in our lab environment

To merge the different cues as well as to deal with uncertainties, we make use of a Bayesian framework. The advantage of Bayesian reasoning is that it allows:

- making explicit statements about the relevance of properties for a certain object,
- introduction of learning in terms of prior and conditional probabilities, and
- assessing the relative importance of each type of relation for the detection of a given object, using the conditional probabilities.

Bayes formula (e.g., see [17]) enables to infer the probability of an unknown event conditioned to other observable events and to prior likelihoods. Let $P(e_i^{\Pi})$ be the prior probability of the occurrence of an event e_i^{Π} (e.g., the probability that any primitive lies in the ground plane). Then, $P(e_i^{\Pi}|\Pi \in \mathcal{O})$ is the conditional probability of the visual event e_i given an object $\mathcal{O}$.

Our aim is to compute the likelihood of a primitive Π being part of an object $\mathcal{O}$ given a number of visual events relating to the primitive:

$$P(\Pi \in \mathcal{O}|e_1^{\Pi},\ldots,e_n^{\Pi}). \tag{1}$$

According to Bayes formula, equation 1 can be expanded to:

$$\frac{P(e_1^{\Pi},\ldots,e_n^{\Pi}|\Pi \in \mathcal{O})P(\Pi \in \mathcal{O})}{P(e_1^{\Pi},\ldots,e_n^{\Pi}|\Pi \in \mathcal{O})P(\Pi \in \mathcal{O}) + P(e_1^{\Pi},\ldots,e_n^{\Pi}|\Pi \neg \in \mathcal{O})P(\Pi \neg \in \mathcal{O})}. \tag{2}$$

In this work we assume independence between $e_1^{\Pi},\ldots,e_n^{\Pi}$ (we intend to investigate to what degree this assumption holds in a future work). If $e_1^{\Pi},\ldots,e_n^{\Pi}$ are independent then $P(e_1^{\Pi},\ldots,e_n^{\Pi}|\Pi \in \mathcal{O})$ can be written as:

$$P(e_1^{\Pi},\ldots,e_n^{\Pi}|\Pi \in \mathcal{O}) = P(e_1^{\Pi}|\Pi \in \mathcal{O}) \cdot \ldots \cdot P(e_n^{\Pi})|\Pi \in \mathcal{O}), \tag{3}$$

and

$$P(e_1^{\Pi},\ldots,e_n^{\Pi}|\Pi \neg \in \mathcal{O}) = P(e_1^{\Pi}|\Pi \neg \in \mathcal{O}) \cdot \ldots \cdot P(e_n^{\Pi}|\Pi \neg \in \mathcal{O}), \tag{4}$$

and the formula (2) becomes rather easy.

Using this framework for detecting lanes, we first need to compute prior probabilities. This is done by hand selecting the 3D primitives being part of a lane in a range of scenes and calculating the relevant relations for these selections. The

Table 1. Prior probabilities

Type	Probability
$P(\Pi\ in\ lane)$	0.44792
$P(\Pi\ not\ in\ lane)$	0.55208
$P(\Pi\ being\ black)$	0.70058
$P(\Pi\ being\ black\ \|\ \Pi\ in\ lane)$	0.97959
$P(\Pi\ being\ black\ \|\ \Pi\ not\ in\ lane)$	0.47391
$P(\Pi\ in\ ground\ plane)$	0.49925
$P(\Pi\ in\ ground\ plane\ \|\ \Pi\ in\ lane)$	0.95960
$P(\Pi\ in\ ground\ plane\ \|\ \Pi\ not\ in\ lane)$	0.12543
$P(\Pi\ has\ normal\ distance\ d_{far})$	0.35943
$P(\Pi\ has\ normal\ distance\ d_{far}\ \|\ \Pi\ in\ lane)$	0.66433
$P(\Pi\ has\ normal\ distance\ d_{far}\ \|\ \Pi\ not\ in\ lane)$	0.11131
$P(\Pi\ has\ normal\ distance\ d_{near})$	0.41015
$P(\Pi\ has\ normal\ distance\ d_{near}\ \|\ \Pi\ in\ lane)$	0.86170
$P(\Pi\ has\ normal\ distance\ d_{near}\ \|\ \Pi\ not\ in\ lane)$	0.04377

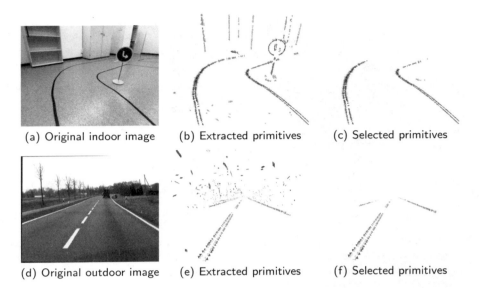

(a) Original indoor image   (b) Extracted primitives   (c) Selected primitives

(d) Original outdoor image   (e) Extracted primitives   (f) Selected primitives

Fig. 8. Extracting the lane in two scenarios: (a-c) showing our indoor lab environment and (d-f) showing an outdoor scenario

results are shown in table 1. The numbers reveal that 'being in ground plane' and 'near normal distance' are the strongest relations as they show the largest difference in probability between the conditions 'in lane' and 'not in lane'.

Figure 8 shows the results of using the Bayesian framework with the computed prior probabilities in two different scenarios: our indoor lab environment and an outdoor scene. The same prior probabilities were used in both scenarios, but for the outdoor scene, the values and thresholds of the relations underlying the probabilities had to be changed to fit the color and dimensions of a real lane.

4.2   Associating Actions to Co-planar Groups

To underline the embedding and strength of our approach of utilizing semantic relations between visual events in the hierarchical representation described in section 2, we briefly present new results on an application that has been described in more detail in [1]. In this application, relations between primitives (or groups) become associated to actions. In figure 9 (left bottom), a grasping hypothesis connected to a co-planar pair of primitives is shown. Hence, the co-planarity graph shown in figure 9 (right), corresponding to the white butter dish, can be associated to grasping hypotheses (as indicated in the middle of the figure). In [18], we could show that by such a simple mechanism, objects in rather complex scenes can be grasped with a high success rate. In figure 10 (left), a scene with a number of objects is shown. Using the grasping reflex described in 9, it was possible to clean the scene (after approximately 30 grasping attempts) except

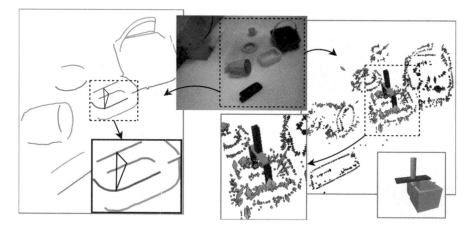

Fig. 9. The 2D contours extracted from the example view on the top-middle are drawn in different colors on the left. The coplanarity graph of the white cup is also shown in black on the left, and this graph suggests a grasp of the type shown in the lower right (the red spheres represent two coplanar representative primitives out of the two contours). The resulting grasp is shown on the left and in the bottom-middle image.

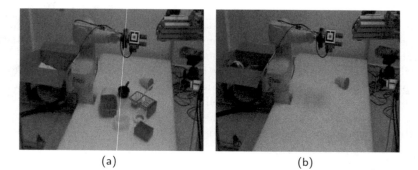

(a)   (b)

Fig. 10. Co-planar pairs of contours predict groups. (a) The four different elementary grasping actions defined based on a pair of co-planar groups. (b) Robot scene before the grasping procedure has been applied. (c) Scene after all graspable Objects have been removed by the system.

one object for which the system's embodiment precluded grasping (i.e., the two finger grasper attachment of the robot could not grasp the round can in any way).

5   Discussion

In this work, we introduced a hierarchical representation of semantically rich descriptors and their relations, and argued that labeled graphs are a suitable framework for scene representation, enabling cue merging and action association. Within this representation, Bayesian reasoning has been applied for efficient cue-merging, allowing for relating textual descriptions to extracted visual information. We also outlined that in such a framework feedback mechanisms at different levels can be used to disambiguate the information, in particular through feedback between the symbolic and signal level.

In our current work, we are aiming at the development of efficient matching strategies that realize the full potential of our representations. In particular, we are interested in structures that cannot be completely defined by their appearance only (as for example in the case of street signs) but by the relations of sub-structures to each other (as, for example, in case of the task of distinguishing different kinds of road structures such a motorways, crossings, motorway exits but also in other more general object categorization tasks).

Acknowledgements

This work has been supported by the European Commission - FP6 Project DRIVSCO (IST-016276-2).

References

1. Aarno, D., Sommerfeld, J., Kragic, D., Pugeault, N., Kalkan, S., Wörgötter, F., Kraft, D., Krüger, N.: Early reactive grasping with second order 3D feature relations. In: Lee, S., Suh, I.H., Kim, M.S. (eds.) Recent Progress in Robotics; ViableRobotic Service to Human, ICAR 2007. LNCIS. Springer, Heidelberg (2007)
2. Echtler, F., Huber, M., Pustka, D., Keitler, P., Klinker, G.: Splitting the scene graph – using spatial relationship graphs instead of scene graphs in augmented reality. In: GRAPP 2008: Int. Conference on Computer Graphics Theory and Applications (2008)
3. Jaromczyk, J.W., Toussaint, G.T.: Relative neighborhood graphs and their relatives. Proceedings of the IEEE 80(9), 1502–1517 (1992)
4. Mucke, E.P.: Shapes and implementations in three-dimensional geometry. Technical report, University of Illinois at Urbana-Champaign, Champaign, IL, USA (1993)
5. Korting, T.S., Fonseca, L.M.G., Dutra, L.V., da Silva, F.C.: Image re-segmentation – a new approach applied to urban imagery. In: VISAPP 2008: Int. Conference on Computer Vision Theory and Applications (2008)
6. Tremeau, A., Colantoni, P.: Regions adjacency graph applied to color image segmentation. IEEE Transactions on Image Processing 9(4), 735–744 (2000)
7. Hancock, E.R., Wilson, R.C.: Graph-based methods for vision: A yorkist manifesto. In: Proc. of the Joint IAPR International Workshop on Structural, Syntactic, and Statistical Pattern Recognition, London, UK, pp. 31–46. Springer, Heidelberg (2002)
8. Kovesi, P.: Image features from phase congruency. Videre: Journal of Computer Vision Research 1(3), 1–26 (1999)
9. Nagel, H.H.: On the estimation of optic flow: Relations between different approaches and some new results. Artificial Intelligence 33, 299–324 (1987)

10. Sabatini, S.P., Gastaldi, G., Solari, F., Diaz, J., Ros, E., Pauwels, K., Hulle, K.M.M.V., Pugeault, N., Krüger, N.: Compact and accurate early vision processing in the harmonic space. In: International Conference on Computer Vision Theory and Applications (VISAPP) (2007)
11. Felsberg, M., Sommer, G.: The monogenic signal. IEEE Transactions on Signal Processing 49(12), 3136–3144 (2001)
12. Krüger, N., Lappe, M., Wörgötter, F.: Biologically motivated multi-modal processing of visual primitives. Interdisciplinary Journal of Artificial Intelligence & the Simulation of Behaviour, AISB Journal 1(5), 417–427 (2004)
13. Pugeault, N.: Early Cognitive Vision: Feedback Mechanisms for the Disambiguation of Early Visual Representation. Ph.D thesis, Informatics Institute, University of Göttingen (2008)
14. Kalkan, S., Pugeault, N., Krüger, N.: Perceptual operations and relations between 2d or 3d visual entities. Technical Report 2007-3, Technical report of the Robotics Group, Maersk Institute, University of Southern Denmark (2007)
15. Piegl, L., Tiller, W.: The NURBS book, 2nd edn. Springer, New York (1997)
16. Kalkan, S., Yan, S., Krüger, V., Wörgötter, F., Krüger, N.: A signal-symbol loop mechanism for enhanced edge extraction. In: VISAPP 2008: Int. Conference on Computer Vision Theory and Applications (2008)
17. Pearl, J.: Probabilistic Reasoning in Intelligent Systems: Networks of Plausible Inference. Morgan Kaufmann Publishers, Inc., San Francisco (1988)
18. Popović, M.: An early grasping reflex in a cognitive robot vision system. Master's thesis, University of Southern Denmark (2008)

Object Detection for a Humanoid Robot Using a Probabilistic Global Workspace

Andreas Fidjeland, Murray Shanahan, and Alexandros Bouganis

Department of Computing
Imperial College London
180 Queen's Gate
London SW7 2AZ, UK

Abstract. We present an architecture for detecting generic objects in unstructured scenes for an embodied visual system. The proposed architecture integrates the contributions of a collection of loosely coupled processes, each supplying a different type of information derived from a robot's sensors, including vision and kinesthesia. The core of the architecture is a probabilistic global workspace, which is used to incrementally build a representation of the scene, and whose contents are made available to the whole cohort of processes. The loosely coupled nature of the architecture facilitates parallelisation, and makes it easy to incorporate additional processes providing new sources of information. We provide an instantiation of this architecture using five processes on an upper-torso humanoid robot. Preliminary results show that the system can classify the elements of a scene well enough for the robot to be able to detect and touch a variety of movable objects within its reach.

Keywords: cue integration, vision systems architecture, global workspace.

1 Introduction

Truly general object detection is still an unsolved problem. Highly specialised methods have been developed for a number of vision sub-problems, such as for stereopsis and optical flow. However, combining these methods into scalable, robust object-detecting systems proves challenging.

The present paper describes a prototype vision system for domain-general object detection based on the idea of combining (or "fusing") the contributions of multiple loosely coupled information "channels" into a small set of probabilistic maps. These maps combine information over time from multiple channels and from multiple sensory modalities. The principle underlying the work is that a large number of individually weak cues, supplied by different channels, can be combined to generate high quality maps of the scene, even when the scene is unstructured and lighting conditions are poor, as they typically are in real life. The set of maps are in turn made available to the full cohort of channels. Each channel can thus take advantage of the earlier contributions from the other

B. Caputo and M. Vincze (Eds.): ICVW 2008, LNCS 5329, pp. 135–148, 2008.

channels, but are only loosely coupled to these through the shared data structure. The architecture is somewhat inspired by the work of Baars [1]. In his terms, the set of probabilistic maps can be likened to a global workspace (or blackboard), while the information channels can be likened to a set of processes competing to influence the global workspace.

The channels include standard and quite general low-level vision routines, such as region detection, stereo matching, and motion detection. The vision problem is best considered in the context of an embodied agent. Some channels thus encapsulate knowledge specific to this agent, such as its kinematics, and use this to update the global workspace. With a wide range of channels, one subset of the available channels might supply the most informative cues under certain conditions (a uniform background, say) while a different subset of channels might supply the most informative cues under a different set of conditions (prominent motion, for example). Our architecture ensures that the most informative channels make the greatest contribution to the overall representation of the scene, and manages these transitions seamlessly.

To achieve robust results, we aim to make use of a large number of channels to cater for a diverse range of situations. Scalability is therefore a primary concern and is achieved through the loose coupling inherent in our architecture. The integration of channel outputs is based on probability theory, but we make use of independence assumptions, where appropriate, to simplify the fusion process. The loose coupling also facilitates the easy addition of new channels, enabling us to incorporate suitable off-the-shelf vision algorithms.

The experimental set-up used to test our prototype implementation was an upper-torso humanoid robot with two 3-degree-of-freedom arms and a stereo camera mounted on a pan-and-tilt head (Fig. 1). The chosen test task was

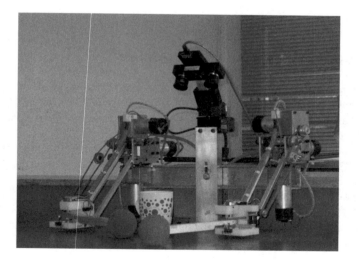

Fig. 1. LUDWIG, a humanoid upper-torso robot with two 3-degrees-of-freedom arms and a stereo camera mounted on a pan-and-tilt head

visually to detect movable objects placed on the robot's workbench, and then to touch them using visual servoing. To achieve this, we use a global workspace containing a depth map and a scene label map which classifies parts of the scene as either background, work surface, robot body, or manipulable object. These labels can in turn be related to affordances.

Our main contributions are:

- The formulation of a probabilistic global workspace architecture for combining evidence from a collection of loosely coupled information channels in probabilistic scene label and depth maps (Sect. 3);
- a prototype implementation of this architecture containing five channels based on a diverse range of routines for image and kinesthetic data processing (Sect. 4); and
- an evaluation of the accuracy of the scene classification of the implemented system (Sect. 5).

The architecture is shown to perform well with respect to the chosen task of object detection. The combination of several processes is found to perform better overall than any subset of processes.

2 Related Work

While the proposed system's motivation comes from the global workspace theory from neuroscience [1], some of the general concepts are also found in earlier symbolic blackboard architectures [2,3]. More recently, but in the same vein, Guhl and Shanahan [4] present a blackboard architecture sharing many of our aims. Their architecture has a symbolic foundation where processes create clusters of features on a hierarchically organised blackboard. The hypotheses thus created are evaluated based on a system of heuristically determined confidence and support values. Our work is based on a probabilistic foundation thereby reducing many of the problems associated with heuristic evaluation and also differs by classifying different parts of the scene rather than attempting to directly detect objects.

Hoiem et al. [5] highlight the utility of a rough scene understanding as a prior for more refined processing, which relates to our use of a scene label map. They describe a method for incorporating low-level object detectors, rough 3D scene geometry, and approximate camera position, modelling their relationships in 3D space. The main aim of the modelling is to provide priors for specific object detectors.

Our work is related to image labelling, as we attempt to segment the scene into different parts. Probabilistic methods are popular in this field, such as Markov Random Fields (MRFs) and Conditional Random Fields (CRFs). These are both graphical methods but differ with respect to the relationships they model. CRFs model the global relationships within the image, such as He et al. [6] who use conditional random fields at multiple scales such that global features provide a context for local ones. We attempt to provide such context, but do not directly build these into our probabilistic models. Rather the overall architecture incorporates several maps, which can provide the required context for feature detectors.

Much existing work combine different cues probabilistically such as stereo with colour and contrast [7]; structure-from-motion, object recognition and tracking [8]; or motion, background subtraction, and skin colour cues [9]. These methods along with image labelling methods usually attempt to carefully model the interactions between modules, thus creating strongly coupled systems, typically with only a small number of modules. In contrast we make simplifying assumptions to achieve a weakly coupled system, which should prove more scalable, both in terms of design effort and computational requirements, as we increase the number of modules in our architecture.

Our work is more closely related to the cue integration method of Hayman and Eklundh [10], which uses loosely coupled modules with independence assumptions, with a strong focus on motion and tracking. We make use of more modules and handle static, as well as moving scenes. Additionally, our method is part of an embodied system, rather than dealing with stand-alone image sequences.

Many of these methods from the image processing literature consider individual images in isolation, rather than image sequences in an embodied system, Such an approach ignores much information that is available to a general vision system such as data from other sensor modalities and time-invariants in the image data.

3   Global Workspace Architecture

A global workspace provides a shared data structure on which information channels operate independently. The information channels are modules which are specialised for a particular task, such as detecting a particular feature in sensor data. These channels communicate only via this shared data structure; When executing, they can consider the contents of the global workspace as well as low-level sensor data (Fig. 2). When a channel detects something noteworthy it updates the global workspace, the contents of which are broadcast to the other channels.

The global workspace takes the form of several *maps* which together represent the robot's belief regarding the state of its surroundings. Each map in the global workspace and its corresponding integration method is a cue integrator. The architecture as a whole combines multiple such cue integrators, by making the data from each map available to all channels. While some channels operate mainly on input data, others operate on the contents of one map and write to another map. Updates to one map can thus propagate to other maps. The whole architecture forms a recursive state estimator. We assume that data arrive at discrete time steps, at which the maps are updated. Channels operate on both sensor data (from the current time step) and the global workspace contents (from the previous time step).

For each map we have a number of desiderata: its representation should be probabilistic; it should have a sound method of integrating information from

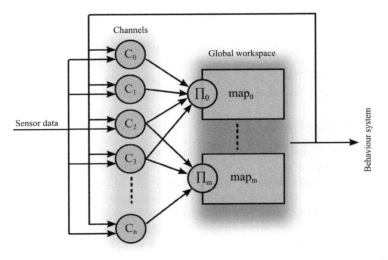

Fig. 2. System overview. The global workspace provides a data structure shared by multiple channels. The channels (C_0–C_n) are specialised modules which take as input both low-level data and the existing contents of the global workspace. The global workspace contains several maps, each with an associated integration mechanism, $\prod$.

multiple sources; and the communication and computation overheads associated with data integration should be low. The first two desiderata ensure that we can represent uncertainty and can deal with conflicting information. The third desideratum is intended to ensure that the architecture is scalable in terms of the number of channels. We plan to extend the system to contain a large number of channels. Furthermore, we wish the architecture to be suitable for fast execution on modern multi-core processors or on reconfigurable hardware. The desired low communication and computation overheads can be achieved by making simplifying independence assumptions in the underlying probabilistic models. We thus delegate much of the complexity of the system to within each channel, rather than to the integration mechanisms.

In addition to the performance benefit of this weak coupling, it also provides a scalable design methodology. New channels can be designed and added to the system without considering dependencies on other channels. The simplifying independence assumptions, however, means that the choice of what channels to add must made with some care.

The robot's surroundings are represented in a head-centred sensory ego-sphere [11], which we call the *scene*. Points in the real world are projected onto a sphere centred on the pan-and-tilt axes intersection at the base of the robot's head. A *scene location* on this map is a coordinate pair specifying pan and tilt angles. This type of map provides a fixed coordinate system onto which events and parts of the scene can be mapped. For each camera position, the image data map onto a part of the ego-sphere.

The global workspace of our current prototype uses two sensory ego-sphere maps: a depth map (Sect. 3.1) and a part-of-scene label map (Sect. 3.2). The depth map provides a 2.5D representation of the surroundings. The part-of-scene label map probabilistically segments the surroundings into background, ground, foreground, and self. We intend to extend the global workspace to also include a motion map and a probabilistic segmentation into distinct surfaces.

3.1 Probabilistic Depth Map

The probabilistic depth map maintains the robot's current belief regarding the depth at each scene location. This map can be updated by several different types of channels, such as depth from stereo, depth from motion, etc. For convenience, the map is expressed in terms of inverse depth (disparity). The belief at each scene location is represented by a univariate Gaussian in its canonical parametrisation $\omega_t = 1/\sigma_t^2$ and $\xi_t = \mu/\sigma_t^2$. For the LUDWIG platform, the map is initialised to a plane below the robot extending to the horizon, with a high variance.

Data from the different channels are integrated at each scene location using an *information filter*. Each scene location is treated independently. Clearly, neighbouring locations are highly dependent, but these dependencies are handled within each contributing channel rather than in the information filter. We furthermore treat contributions from the different channels as independent.

The map update takes place in two stages: prediction and measurement update. First, the prediction update:

$$\bar{\omega}_t = (\omega_{t-1}^{-1} + R)^{-1} \quad \text{and} \quad \bar{\xi}_t = \bar{\omega}_t\, \omega_{t-1}^{-1}\, \xi_{t-1}$$

R is the process uncertainty, and there is no control vector.

A number of channels C may contribute to the depth map. Each of these channels $c \in C$ provides at each time step t a disparity mean $\mu_{c,t}$ and variance $\sigma_{c,t}^2$. Under independence assumptions the information filter update simplifies to

$$\omega_t = \bar{\omega}_t + \sum_{c \in C} \frac{1}{\sigma_{c,t}^2} \quad \text{and} \quad \xi_t = \xi_{t-1} + \sum_{c \in C} \frac{\mu_{c,t}}{\sigma_{c,t}^2}$$

Each channel update now only involves adding terms to two accumulators. This can be done independently of other channels, and in any order.

3.2 Part-of-Scene Label Map

It is useful to segment the scene into different relevant categories. We use the set of mutually exclusive labels $\mathcal{L} = \{$ ground, foreground, self, background $\}$. A scene location is labelled ground if it is part of the robot's work surface. The ground provides support for the manipulable objects we are interested in. A scene location is labelled foreground if it is part of a manipulable object. A scene location is labelled self if it contains any part of the robot's own body. A scene location which does not fall into any of the above categories is labelled

background. This above classification of the scene serves two purposes. First, as the labels relate directly to affordances, the scene classification can be used to direct behaviour. For example, a touching behaviour may direct attention to parts of the scene which are labelled **foreground**. Second, the classification can serve to direct attention and processing power for more refined analysis to the parts of the scene which are relevant to the current task.

The label map maintains the current belief, $p(X_t|z_{0:t})$ at each location as a discrete probability distribution over the labels. X_t is the random variable at time t and $z_{0:t}$ denotes the input, both video and sensor data, up to time t. We model the system as a hidden markov model (HMM) where the scene labels are the hidden states. Similar to the depth map update, each location is treated independently and dependencies between neighbouring locations are handled within each channel.

The observed state z_t is either image or sensor data. Because of its high dimensionality we do not directly provide measurement probabilities of the form $p(z_t|X_t = l)$. Instead each channel c provides an observation probability of the form $p(f_c(z_t)|X_t = l)$, where f_c is a channel-specific feature detector of lower dimensionality. We make the simplifying assumption that these channel outputs are independent and thus define the observation probability as follows:

$$p(z_t|X_t = l) = \prod_{c \in C} p(f_c(z_t)|X_t = l)$$

The channel-specific observation models $p(f_c(z)|X)$ are learned from labelled data. Typically, f_c is a discrete function, so that the observation model is a $4 \times n$ matrix where n is the number of discrete values provided by f_c. We limit the certainty that the observation model can express, to avoid overfitting to the training data. Additionally, we impose constraints on the model based on prior knowledge where we know a priori that the distribution for specific data values should take a certain form. For example, we may know that a channel is able to distinguish one label from all the others, but may not be able to distinguish between these other labels.

To update the belief we use a discrete Bayes filter (HMM forward pass). The predicted belief for label $k \in \mathcal{L}$ is

$$p(X_t = k|z_{0:t-1}) = \bar{p}_{k,t} = \sum_{l \in \mathcal{L}} p(X_t = k|X_{t-1} = l)p_{l,t-1}$$

based on state transition probabilities learned from labelled data. Data from each channel are incorporated to attain the next state estimate

$$p(X_t = k|z_{0:t}) = p_{k,t} = \eta \; p(z_t|X_t = k) \; \bar{p}_{k,t}$$

The normaliser η ensures that the probabilities sum to 1.

Similar to the depth map update, each channel update only involves multiplying terms to an accumulator which can be done independently from other channels and in any order, thus ensuring low coupling.

3.3 Behaviour System

The global workspace architecture is used in the upper-torso humanoid robot
LUDWIG as part of an active vision system. The contents of the global workspace
are used to direct attention and drive behaviours. Both the scene label map and
depth map are useful in this respect. The scene map may be used to direct attention to behaviour-relevant parts of the scene, while the depth map specifies
the 3D locations of those parts.

Our current focus is on directing attention towards foreground objects with
the aim of manipulating them. To this end, we derive an interest map from the
scene label map that indicates the presence of foreground pixels. Each location
on the interest map combines two terms: label interest (i^+), and boredom (i^-).
At each scene location we compute the total interest at time t: $i_t = i_t^+ - i_t^-$. We
find a region of interest (ROI) for the behaviour by finding a fixed-size area in
the interest map in which the sum of i_t is maximal and above some threshold.

The label interest term is high in the parts of the scene which are highly likely
to contain foreground objects:

$$i_t^+ = \begin{cases} p(\texttt{foreground}|z_{0:t}) & \text{if} \quad \texttt{foreground} = \operatorname*{argmax}_{x \in \mathcal{L}} p(x|z_{0:t}) \\ 0 & \text{otherwise} \end{cases}$$

The boredom term ensures that the attention switches from the current ROI
to some other part of the scene after some time.

$$i_t^- = \begin{cases} i_{t-1}^- + w_{\text{ROI}} & \text{if in ROI} \\ i_{t-1}^- - w_{\overline{\text{ROI}}} & \text{if } i_{t-1}^- > 0 \\ i_{t-1}^- & \text{otherwise} \end{cases}$$

where the weight w_{ROI} influences the time a behaviour remains fixated at a
location, and the weight $w_{\overline{\text{ROI}}}$ influences the time before the behaviour can be
directed back to a previous region of interest.

The above method can be used for other behaviours as well, either by providing suitable parameters for existing labels, or by introducing new labels such
as a human label used for a gaze-following behaviour. With multiple behaviours,
certain conditions in the global workspace can trigger transitions between behaviours in an FSM-style behaviour model. The ROI may have a different meaning depending on the behaviour. An exploring behaviour may direct the gaze
towards the region-of-interest; a touching behaviour may additionally move the
hands there.

4 Five-Channel Prototype

The previous section presents a general architecture for combining the output
of specialised channels in order to infer the structure of a scene. We provide a
prototype implementation with five channels. Some of these channels use general
image processing techniques, such as depth from stereopsis, while others are

specific to the LUDWIG robot. Furthermore, some channels operate mainly on input data, updating the global workspace, while some channels use data from one part of the global workspace to update another.

Kinesthesia. The robot is equipped with rotation sensors on the arm joints. The *kinesthesia* channel updates both the depth map and the scene label map to take into account the current position of the arms. The channel segments the scene into self, ground over which the arms have recently moved ("moved-over"), and non-self. The "self" parts of the scene are found by ray-casting onto a sphere-and-cylinder model of the arms. We keep track of the areas over which the arms have moved, as this is a strong indicator that this part of the scene is ground with no objects on it. The channel updates the depth map for pixels labelled as "self" or "moved-over". For "self" scene locations the channel uses the depth found by ray casting as the mean with a small fixed variance. For "moved-over" scene locations, the mean depth is set to the distance to a horizontal plane at the base at the robot. The *kinesthesia* channel can distinguish self from the other labels but does not distinguish between these other labels. The model learned under this constraint specify that self is significantly more likely than the other labels at the scene locations where the arms are currently believed to be located. For the "moved-over" parts of the scene, ground is more likely than foreground, and both self and background are highly unlikely.

Stereopsis. The *stereopsis* channel estimates the scene depth based on short baseline stereo. We use a dynamic programming stereo algorithm due to Birchfield and Tomasi [12], operating on pairs of Sobel edge maps.

Reachable. The *reachable* channel connects the depth map and the scene label map. The channel takes into account the limited reach of the robot, and segments the scene into reachable and unreachable parts based on the depth map in the global workspace. The depth threshold is derived using the dimensions and kinematics of the robot. Scene locations within reach can be more readily explained as foreground, ground, and self, and less readily explained as background.

Uniform Surfaces. Since real-world objects classified as foreground or self are naturally limited in size, a large uniform surface would normally be ground or background. To detect such uniform surfaces we use a sparse circular mask around each pixel, and compute a similarity measure based on the difference in hue and saturation between the central pixel and each of the surrounding pixels. If the central pixel is part of a large uniform surface, most surrounding pixels will be similar in hue and saturation, which leads to a strong response in the mask. This detector provides a fast method of directly detecting large surfaces, without doing a full segmentation. We learn an observation model for this channel based on ground-truth labelled data. In the learned model, a strong response (high similarity to surroundings) is deemed likely to be caused by a true ground or background label. A weak response (low similarity to surroundings) does not provide much discriminatory power between the labels. Such weak responses are typically found for true foreground and self labels, as well as for coarsely textured `ground` and `background` surfaces. This channel thus mainly serves to reject false positives for `foreground` and `self`. The method could also be applied using similarity measures based on texture rather than colour.

Differential Motion. The *differential motion* channel uses differential motion analysis to determine what parts of the scene contains moving objects. It performs segmentation based on an accumulated history of image difference, and produces a binary map of changing and non-changing parts of the scene. Image changes are assumed to be caused by motion, while other causes of image change is considered noise. The channel considers the previous state of the scene labelling. For example, if a location at which motion was detected was previously believed to be `background`, motion is more readily explained by the presence of `self` or `foreground` objects. If, on the other hand, a location was previously labelled `foreground`, motion is more readily explained by `foreground` (moving textured object) or `background` (object moved away, revealing background). The channel learns the model from labelled data.

The five channels presented above are only small sample of the possible channels that can be brought to bear on the problem, but nonetheless provide a useful prototype for evaluating our architecture. As we extend the system other channels can be added, for example considering optical flow, texture, shadows, projective geometry, lighting, face detection, and depth from motion.

5   Experimental Results

To evaluate the performance of our architecture using the five channels described in Sect. 4, we conduct experiments using the LUDWIG robot in different scenarios. We evaluate the quality of the labelling as well as evaluate the effect of using different channel combinations.

5.1   Scenarios

The input to the system is a stereo RGB video stream with data at a resolution of 640x480 along with a stream of sensor data for the pan/tilt unit and arm rotation sensors. We run the robot through three different scenarios involving both camera and arm motion, and with varying levels of clutter (Fig. 3). We have hand-labelled ground truth scene labels for each sequence. The observation models used in the evaluation were learned from a different set of ground-truth labelled data.

The three scenarios involve the same set of objects, but we vary the background as well as the interaction with the objects to evaluate the performance of the system under different, but controlled situations. Four objects are present in the scene. The system has no prior knowledge regarding either of these, but must identify these objects based on their observable properties.

In the first scenario (*uniform*), the objects are placed on the uniformly textured wooden workbench on which the robot is mounted. During the sequence

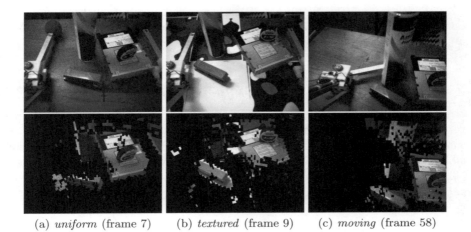

(a) *uniform* (frame 7)   (b) *textured* (frame 9)   (c) *moving* (frame 58)

Fig. 3. Scenarios used in evaluation of system, showing input frame (top) and detected objects (bottom)

the camera moves twice to observe different parts of the scene. In the second scenario (*textured*), the workbench surface is covered by coarsely textured paper before placing the objects, resulting in a significantly more cluttered scene. The camera motion is the same as for *uniform*. In the third scenario (*moving*), we use the same basic setup as for *uniform*, but additionally have the robot touch and push one of the objects (the spray can). This scene thus adds both self motion and object motion.

5.2 Experiments

To evaluate how well the system makes use of the different channels, we run the robot using different channel combinations for each scenario. The combination of all channels is used as the baseline, with which we compare combinations where one channel has been removed. The channel combinations where one channel is removed gives an indication of the contribution of that channel.

We evaluate the performance of the system using ROC curves for the scene labelling. This plots the true positive rate for classification of foreground pixels against the false positive rate at different threshold values for $p(\texttt{foreground}|z_{0:t})$ at which a location is classified as `foreground`. The ROC curve is determined with respect to this label only on a per-frame basis. To get a single measure of the system performance for each frame, we compute the area under the ROC curve. For this area-under-curve (AUC) measurement, a value of 1.0 corresponds to a perfect classifier, whereas a 0.5 corresponds to a random classifier.

5.3 Discussion

The bottom row of Fig. 3 shows the parts of the input images classified as foreground according to the method in Sect. 3.3. There are some mis-classifications

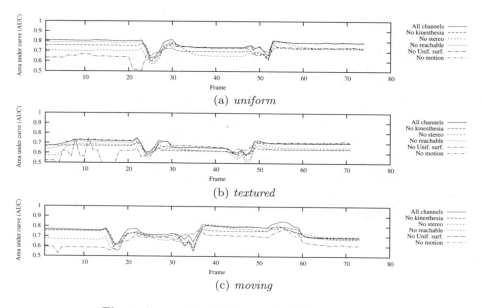

Fig. 4. Area under ROC curve for different scenarios

and the output is somewhat coarse due to being back-propagated from the scene coordinate system, which is lower resolution. Overall, though, it is clear that the main objects in the scene are chosen over the surroundings, also in the cluttered *textured* scenario. Fig. 4 shows the AUC values for each of the video sequences. Within each sequence the performance is fairly constant, even as the camera moves to different parts of the scene, with different characteristics. Each of the sequences has pronounced reductions in performance. These correspond to frames during and shortly after camera motion. The performance is lower in these parts of the sequence as the recursive state estimation takes a few frames to stabilise. The performance differs slightly between the sequences. The simpler *uniform* scenario has the highest overall performance among the three, followed by the *moving* and *textured* scenarios.

Combining channels increases performance. For most frames, leaving one channel out either decreases performance, in some cases dramatically, or has no effect. Although for some frames there is a slight improvement in performance when leaving a channel out, this effect is generally small and short. For no channel does removing it improve the performance overall.

We observe that different channels contribute in different situations. Consider for example the *uniform* scenario. In the first third of the sequence the *uniform surface* channel makes the greatest contribution; removing it has the greatest negative impact. In the middle third of the sequence, however, the stereo channels are more important. In this part of the sequence removing *uniform surface* has almost no effect. These two parts contain different views of the scene, the first mostly containing the workbench, while the second also captures some of the background. In the latter case, the stereo data are important to avoid false classifications. In the final third of this sequence both the *uniform surface* and the *kinesthesia* channel have significant contributions. In this part of the sequence, the robot's left arm is in the centre of the frame and the *kinesthesia* avoids mis-classifying this.

Consider also the *textured* scenario. Because of the coarse texture present in this sequence, detecting uniform surfaces is of little use. Instead, e.g. from frame 50 onwards the *reachable* and *kinesthesia* become the most critical channels in achieving good object detection. A single channel may contribute to short-lived situations. The presence of the *motion* channel makes little difference for most of the sequences. In the *moving* scenario, however, it is the dominating channel in classifying the objects in the motion shortly before frame 58.

We thus observe two traits of the system: an increased overall performance when adding new informative channels, and different channels dominating the object detection under different circumstances. We therefore expect to further improve the performance and robustness of the system by adding new channels.

6   Conclusions and Future Work

The probabilistic global workspace architecture presented in this paper provides an attractive means by which to construct robust vision systems for object detection in challenging scenes. The results from our prototype system show the promise of the approach, with our robot able to direct attention on objects in the scene based on generic clues rather than some specific object model. The modular nature of the architecture enables us to easily add channels as we in future work extend the system with additional channels, which might consider optical flow, texture, shadows, projective geometry, lighting, and depth from motion.

Acknowledgments

The authors gratefully acknowledge the contribution of grant EP/C51050X/1 from EPSRC.

References

1. Baars, B.J.: A Cognitive Theory of Consciousness. Cambridge University Press, Cambridge (1988)
2. Nii, H.P.: The blackboard model of problem solving and the evolution of blackboard architectures. The AI Magazine 7(2), 38–53 (1986)
3. Hanson, A., Riseman, E.: VISIONS: A computer vision system for interpreting scenes. Computer Vision Systems, 303–334 (1978)
4. Guhl, T.P., Shanahan, M.P.: Machine perception using a blackboard architecture. In: Proc. Int. Conf. on Computer Vision Systems, Bielefeld (March 2007)
5. Hoiem, D., Efros, A., Hebert, M.: Putting objects in perspective. In: CVPR (2006)

6. He, X., Zemel, R.S., Carreira-Perpinan, M.A.: Multiscale conditional random fields for image labeling. In: CVPR, vol. 2, pp. 695–702 (2004)
7. Kolmogorov, V., Criminisi, A., Blake, A., Cross, G., Rother, C.: Probabilistic fusion of stereo with color and contrast for bi-layer segmentation. PAMI 28(9), 1480–1492 (2006)
8. Leibe, B., Cornelis, N., Cornelis, K., van Gool, L.: Dynamic 3D scene analysis from a moving vehicle. In: CVPR, pp. 1–8 (2007)
9. Prince, S.J.D., Elde, J.H., Hou, Y., Sizintsev, M., Olevskiy, Y.: Statistical cue integration for foveated wide-field surveillance. In: CVPR, vol. 2, pp. 603–610 (2005)
10. Hayman, E., Eklundh, J.O.: Probabilistic and voting approaches to cue integration for figure-ground segmentation, pp. 469–486 (2002)
11. Albus, J.S.: Outline for a theory of intelligence. IEEE trans. Systems, Man and Cybernetics 21(3), 473–509 (1991)
12. Birchfield, S., Tomasi, C.: Depth discontinuities by pixel-to-pixel stereo. In: ICCV, pp. 1073–1080 (1998)

Author Index

Printing: Mercedes-Druck, Berlin
Binding: Stein+Lehmann, Berlin